DESTINATION SPACE

About the Author

Kenny Kemp is an award-winning journalist and writer. The founding business editor of the *Sunday Herald*, Kenny Kemp was Scotland's Business Writer of the Year in both 2001 and 2003. He is the co-author with Barbara Cassini of *Go: An Airline Adventure*, which won the 2004 Business Book of the Year in the WH Smith Awards, and *Flight of the Titans*. He lives in Edinburgh.

DESTINATION SPACE

Making Science Fiction a Reality

Kenny Kemp

First published in Great Britain in 2007 by
Virgin Books Ltd
Thames Wharf Studios
Rainville Road
London
W6 9HA

A catalogue record for this book is available from the British
Library.

ISBN 978 0 7535 1235 7

The paper used in this book is a natural, recyclable product
made from wood grown in sustainable forests. The
manufacturing process conforms to the regulations of the
country of origin.

Typeset by TW Typesetting, Plymouth, Devon

Printed and bound in Great Britain by
MPG Books Ltd, Bodmin, Cornwall

CONTENTS

FOREWORD BY SIR RICHARD BRANSON

It is remarkable to think that we are on the verge of another new era of space travel and one which will be wholly different to anything that has gone before. My dream is to make space accessible to tens of thousands of people. Enabling future generations to observe the magnificent beauty of our fragile planet and experience weightlessness without the environmental consequences of current ground-based rockets, which were designed during the Cold War, is fundamentally important.

I am often asked, How does going into space square up with the Virgin Group's increasing passion for the environment and green business?

Recently Al Gore came to see me in London and convinced me that a cycle of global warming could lead to the destruction of human civilization on Earth. It has spurred us into action, with Virgin's commitment to invest in bio-fuels and other renewable energy sources over the next ten years. I now believe that we must explore space to find new resources. The survival of humankind may well depend on this. But we have to walk before we can run, and Virgin Galactic's space flights are a good starting point. Our new craft, SpaceShipTwo, designed by the legendary Burt Rutan, will be a technological breakthrough. It can carry either six passengers or the equivalent weight of payload into space for the same CO_2 output as one business-class passenger flying from London to New York.

SpaceShipTwo is truly innovative. It is made from lightweight composite materials, and it will be launched by an inherently safe system from a mothership at 50,000 feet. On return from space there is a simple method of re-entry into the atmosphere, using Burt Rutan's amazing shuttlecock feathering mechanism and then a single glide home to a conventional runway.

So we are on the verge of a revolution in space tourism and as head of the Virgin Group I am proud to be playing my part. Already we have nearly 200 Virgin Galactic astronauts preparing for their flights. People such as scientist Stephen Hawking and actress Victoria Principal are among those keen to become space ambassadors. And there will be thousands to follow them. Fewer than 500 people have ever experienced real space – and we hope to double that number in our first year of operation.

Other new space companies are also approaching this burgeoning market – but for Virgin Galactic there are two fundamental points: safety and the environment. Our revolutionary spacecraft will be inherently safe and a much cleaner and greener way to gain access to space. And while the price of flying those early pioneers seems high to many, we are committed to bringing the prices down as soon as we are happy with the system we are building.

The new space race is for commercial companies bidding to take people and payload into space. In *Destination Space*, Kenny Kemp has splendidly captured the riveting story of how space tourism and commercial space transport is developing around us out of a fifty-year-old government monopoly.

Space is infinite. And it will take many generations to overcome all the hazards and dangers for humans to exist for long periods of time in outer space, so that we can go beyond our solar system. But I believe the indomitable human spirit will play a part in this exploration. This is only the beginning of an incredible journey.

I have reserved my place and I'm extremely excited about the prospect of taking a Virgin Galactic trip into space. I'll be flying with my mum and dad, my daughter, Holly, and my son, Sam, on the inaugural commercial flight within the next three years.

Sir Richard Branson

PROLOGUE

Nearing the end of the writing of this book, I was taking time out to reacquaint myself with my kids. I was sitting in a Starbucks café in an Edinburgh bookshop with my thirteen-year-old daughter Florence, an avid reader who had just finished the last Lemony Snicket adventure.

After her hot chocolate drink with its marshmallow twizzle, she asked, 'How's your book getting on, Dad?'

'Great,' I lied, still wrestling with the story.

'Maybe I'll read it,' she ventured.

'So would you like to go into space?' I inquired.

'I'll go once a few other people have gone and survived. But I think it would be pretty cool.'

This book is for Florence and her generation of space cadets. As Buzz Lightyear says: 'To Infinity, and Beyond!'

1. SIMPLY OUT OF THIS WORLD

P eace. Perfect peace. For a few split seconds there is an unworldly stillness – after the hellish cacophony and skull-crushing vibration that has just pounded your human frame. Now there is only peace. No sounds. No noise. Just a gentle drifting, a hundred kilometres above the blue-rimmed Earth. There is a vague and ethereal whiff of coolness as your limbs are lifted upwards in the pressurised cabin. Your body feels free of its burden and gently you rise, restrained only by the firmness of your seat straps.

This is the out-of-this-world sensation for the suborbital space tourist. It is the culmination of years of planning, preparation and testing – a revolutionary space technology that will take paying passengers out into space. It is the beginning of an infinite human journey that will take civilians out to explore the solar system and the stars beyond. If this all sounds like some crazy science fiction, it is not. This is now commercial reality.

*   *   *

Inside SpaceShipTwo – launched as the Virgin Galactic Spaceship *Enterprise* – there are animated grins of relief. After a bout of terrifying and tempestuous buffeting during the launch into space, each face is now lit up with extraordinary elation.

'You may now unfasten your seat belts,' says a calm voice in your earpiece.

With a click of the button on your chest strap, you are free to float in the cabin. Your body rises, your limbs feel light and unconfined. Around you, fellow passengers are laughing mildly in disbelief at this weird sensation. You reach up and pull yourself to a round porthole above your head. With ease, you glide upwards to look outside. You are suspended in a drifting life raft looking out on the endless inky ocean of the universe. This is the Virgin Galactic experience. It is the ultimate high.

From your porthole, the Gulf of Mexico is laid out like a vast relief map. Now you have joined the elite. You are one of the most privileged human beings in the world. A mere mortal, able to look down on a fragile Earth as weightlessness takes you into a realm experienced by only a very few.

The intercom clicks in your ear. 'Wow! How about that, then? Everyone OK back there?' inquires Alex Tai, Virgin Galactic's chief pilot and now a veteran of takeoff and landings with SpaceShipTwo.

'Unbelievable. Absolutely unbelievable,' replies founder astronaut, Trevor Beattie, who is among the first paying passengers to sign up for the programme.

'Take a look out on the port side,' says Tai. 'That's the Caribbean Sea down there and you can see the South American rainforest too. Looks like a lot of forest fires going on.' You'll notice a thin pall of smoke rising from a dark green blot on the horizon.

In the seat next to you, an attractive blonde woman is strangely subdued as she rises into the air. Moments earlier she had been shrieking in utter panic as her cheeks convulsed with intense pressure, her head beaded in sweat. Now she is the very essence of serenity.

'My God, it all looks so fragile. The horizon is so thin. Is that really where we live?' she asks, rising up gently to her nearby porthole.

'Sure is, and it's the only home we have right now, so we need to get better at looking after it,' replies Tai.

SpaceShipTwo is floating gently. The craft is calm, and as it shifts round towards the sun a stream of brilliant, dazzling white light pours through one of the rounded windows in the roof. For the six astronauts on board, this is worth every penny, dollar or euro they have had to fork out – and more.

But why are people now willing to pay to go into space? Fundamentally, we are an inquisitive species. Captain Kirk was right – space is the final frontier. The possibility of travelling into space helps satisfy an innate instinct: the human impulse to explore. Now this unexplored territory is at last opening up for the traveller. And the prospect of being able to reach out for the solar system, the galaxy and beyond fills us with a deep sense of awe. The Earth is simply a minuscule green and blue microdot in an infinite universe. As the Apollo astronauts will testify, there is something deeply spiritual about seeing how fragile and insignificant our planet seems in the grand scheme of swirling nebulas and supernovas. And more people want to find this out for themselves.

Right now the commercial option is the suborbital flight. Not orbital. An orbit is a moving path around another body – and going around the world in orbit will require much more velocity and energy. But suborbital flight will still give the tourist a genuine experience of space. And it will qualify the paying customer as a real, live, official astronaut. Yes, you can wear the badge and the T-shirt to prove it. Fewer than 500 human beings have ever been into space. The founding Virgin Galactic astronauts will join an elite of brave human beings. If you pay your money and fly, you can claim your place alongside Yuri Gagarin, Neil Armstrong, Buzz Aldrin, Valentina Tereshkova (the first woman in space), Dennis Tito (the first paying passenger) and, more recently, Anousheh Ansari, the first Iranian woman in space.

However, being one of the first has transitory value and limited cachet. Within the next ten years it will become commonplace and your claim to fame as an astronaut might be less newsworthy – the first Milton Keynes lorry driver in space or the first washroom attendant from Mohawk Falls is unlikely to make the national news.

So while there is kudos in being in the very first batch of space tourists – and in an ephemeral world dominated by vacuous celebrity, there will be insatiable interest around the globe – it is not the core reason. The new space-tourism players have signed contracts for reality shows set in space. Contestants can win a chance to go suborbital and have every move documented, and there will even be contests to find astronaut pilots with the right stuff. And, as you will find out, there is keen competition among those who have signed up already to be on the very first Virgin Galactic flights, scheduled for early 2009.

But space is more than just another destination for some lurid and gratuitous entertainment. For an increasing number of people, it is being viewed as the best (perhaps the only) long-term option for the survival of the human race. Nothing less. They argue that our increasingly crowded and polluted world does not have enough resources to feed, heat and clothe our booming population, and global warming is threatening the fragile ecosystems and habitats for so many plants and animals. Space offers the opportunity of finding energy sources and minerals to sustain life, and now physicists, engineers, chemists and geologists face the ultimate challenge – of taking humans into outer space and colonising far-off planets. Now people believe science fiction can become a reality.

The tipping point for the remarkable interest in commercial space travel was the Ansari X Prize. On 4 October 2004, the legendary airplane designer Burt Rutan and financier and Microsoft founder Paul Allen clinched the crown with the flight of SpaceShipOne. The $10 million prize had a simple requirement: to carry the weight of three people 100 km above the Earth's surface, twice within two weeks. The success of this tiny spacecraft has now opened the gates for suborbital flights.

Burt is a genius, pure and simple. Some have called him a 21st-century Leonardo Da Vinci, though the Californian guffaws at such an outrageous suggestion. But there is no doubt he is a central figure in this story. Long before winning this prize, Sir Richard Branson had come to pay homage at Rutan's shrine in the Mojave Desert in California. Now, Virgin Galactic was teaming up with Burt's company, Scaled Composites, for a most incredible project.

The arrival of SpaceShipTwo, a revolutionary flying machine, is making suborbital space travel a reality. It will soon begin a series of test flights, with the aim that passengers chasing their dreams of experiencing space can enjoy the reality within the next two years.

Space has become interesting again. And so much more fun. The Cold War rivalry of NASA and the Soviet cosmonauts was highly political and intensely serious. What is happening now is a space renaissance – and any renaissance is characterised by more colour and flair and a flowering of fresh thinking and activities. This has unlocked the door for interested entrepreneurs and business people who have made their money in fields outside of space.

Along with Sir Richard Branson, there are other high rollers – and trailblazers of the Game Boy generation – anxious to take a punt on commercial space: there is Jeff Bezos, who made billions selling books and other goods in cyberspace with Amazon.com; Paul Allen, the co-founder of Microsoft; the Las Vegas hotel magnate Robert Bigelow, who has launched a large inflatable space-hotel; John Carmack, the computer-games creator behind teenage hits such as Doom and Quake; Elon Musk, the founder of PayPal, who has set up SpaceX; the Ansari family, who put up the $10 million for the X Prize; and Californian Jim Benson, of Benson Space Company, who invented full-text searching and indexing in 1994 with his business called Compusearch and then Imagefast, and sold them out and retired at fifty. They are all adding character and vision to an industry once dominated by earnest NASA officials with crew cuts and immaculately ironed, button-down short-sleeved shirts. Most share a common upbringing – they were awestruck kids when the Apollo Moon landings happened in the late 1960s. And they know that not a single human has left Earth's gravitational pull since 1972.

In 2009, paying passengers will be able to ride into space on a Virgin Galactic mission. They will be able to experience weightlessness, witness the curvature of the Earth and have a view of the universe seen only by astronauts.

So just how high does the space tourist need to go to experience the marvel of the Earth? Space scientists know that 115 miles high is the minimum altitude at which objects are placed into orbit. But, at this height, the velocity required to make it into orbit is roughly 30,100 ft per second, or 20,500 mph. SpaceShipTwo is not designed to achieve this kind of speed and won't be able to carry enough propellant to push it into orbit. That is another step for much later. But there are no arguments that you are now in space, well beyond the 62 miles (100 km) set by the Ansari X Prize competition for sending civilians into space.

Suborbital space is simply the foundation for the space tourist – as new technology comes of age paying passengers and commercial spacecraft will go further and faster, reaching out to the Moon, Mars and beyond. This is the story so far – but there is a universe to explore.

2. TRUTH OR CONSEQUENCES

NEW MEXICO, 2010

The ethanol-fuelled V-Buzz purrs along the baking tarmac of the Interstate 25 freeway. Inside the air-cooled vehicle a dozen expectant passengers peer out from the full-length tinted windows at the unfolding desert landscape. An hour earlier they had embarked from a Virgin Atlantic flight at El Paso on the border of Texas and Mexico. Without fuss or delay, they were cleared through this charming airport, met by their reps and were soon heading north. On the left side are the jagged peaks of the Sierra del Olvido – the Organ Mountains – which soon give way as the road rolls along parallel to an upper Rio Grande, not so grand but able to give some much-needed green to the land.

Two large billboards scream into view. The first is a brashly coloured advertising hoarding with the cutout of a single-seat jet zooming out of the top. 'Rocket Racing League: Next meeting at Las Cruces International Airport. October 16. 2pm. Come See All Your Grand Prix Stars,' it implores.

Then a more prosaic sign appears in blue: 'For the Spaceport America: Virgin Galactic Terminal One, Hotel and Hot Springs Spa: 15 miles. Take Junction 34.'

A few miles later, on the right-hand side, just before the appropriately named Upham, the Virgin passengers see a large and intriguing structure ahead in the heat haze. From the road it resembles a brown sand dune, gleaming in the sunshine. It is almost alien – after all this is New Mexico, home of the Roswell mysteries. Up in the crystal-clear blue sky, a sleek, futuristic aircraft with long, thin wings turns elegant free-flowing circles as it descends towards the building.

'Ah, look, there's a SpaceShipTwo coming back into land. That's us in a few days,' says a 65-year-old English gent.

'Rather you than me, dear,' replies his wife.

As the V-Buzz turns off the main highway another sign, this time in polished stone dressed with Comanche symbols, reads: 'Welcome to the Eighth Wonder of the World: Spaceport America.'

The vehicle pulls off and drives past row after row of low-slung pueblo-style offices. This is a special place where landmark architecture blends in with the orange earth, mesquite bushes and soap-tree yuccas. There is not the conspicuous consumption or the ersatz edifices of Las Vegas – this is New Mexico, an arid mountainous state with a rustic Hispanic history. It's called the Land of Enchantment for good reasons.

Every environmental consideration has moulded the cre-ation of Spaceport America. It looks futuristic and it faces straight up to the stars. The passengers step off the bus and walk into a luxurious oasis. The Virgin Galactic welcoming party is full of smiles and handshakes as everyone is ushered into a cool, high-vaulted, stone-clad chamber with a large circular window in the roof. On a raised platform there are three large telescopes trained at the ceiling. A gentle waterfall laps down into a tempting pond where flowering desert lilies give off a gentle fragrance.

Dr Julia Tizard, Virgin Galactic's director of space medi-cine, is among the welcoming party. An Englishwoman in her early thirties, she is now an acknowledged authority on space tourism.

'Good evening and welcome. Let me first say that what you are about to experience will undoubtedly be one of the highlights of your life. We are here to ensure that you can enjoy every moment, and that you have time to cherish something that will be out of this world.'

This is the start of three days' training – with some fun and local colour thrown in. This is where your out-of-this-world experience begins.

But let's rewind here a few years, before we get too carried away. While space tourism is now with us, it will be several years before Spaceport America is open for the ultimate experience. But New Mexico is now investing hundreds of millions of dollars in this incipient sector of the global tourism industry. This vast American state – half the size of France but with barely 1.8 million of the US's 300 million souls – has decided it wants to be at the forefront of what it sees as an explosion of interest in commercial space travel. Around the world, more people are waking up to the astounding opportunities that will come with commercial space flight.

These are only the first teetering steps of this infant industry. To reach adulthood – and maturity – a lot of technical hurdles still need to be overcome. And, for now, the Spaceport America site is no more than a few concrete foundations and some surveying marker posts in one of the most desolate parts of America. So perhaps it's time for a little reality check. The nearest settlements to the new spaceport, due for completion in 2010, is the city of Truth or Consequences.

It will be in its administrative jurisdiction and this will mean changes to this friendly little place, that appears half asleep even in the height of the tourist season. 'City' is rather a grand word for this town of two main streets running over a hot-spring aquifer where Geronimo and his Indian braves came to soak their wounds after battles. But it is a city. Here is the authentic New Mexican melting pot of Anglo, Hispanic and Native American living in a community of about 8,000 people that is 4,260 ft up in Sierra County. Inside the Cuchillo Café, with the white tablecloths protected by thick transparent sheets of plastic, they still serve home-made lemonade and iced tea, and the Mexican food is wholesome. The Cuchillo Platter-El Grande, a mountain of meat enchilada, cheese enchilada, tamale, taquito and sopaipilla, comes in at $12.25. It's enough to feed a Mexican raiding party. The Geronimo Burger is $6.95.

Orlando Romero, wearing a black Stetson and sporting a grand handlebar moustache, is the owner and proud of his New Mexican heritage. He inherited the café from his grandfather, Felix Sanchez. Orlando worked in computers before retiring to help in the restaurant. 'I've been reading a lot about Richard Branson. He's going to be investing a lot of money in this area – and we could do with some of that,' he says. 'This has always been a town that has welcomed its visitors. We won't mind if there is a spaceport nearby. Already I can see trade in the town is picking up.'

Truth or Consequences was once called Hot Springs, named after the geothermal waters that bubble up under the thoroughfares. One of the most popular bathing spots is the Indian Springs, owned by Cindy Earthman. It has rooms attached and for $3 a dip and 50c for a towel the Virgin Galactic tourist can try the mineral baths. You can't luxuriate in the water too long; it's a rather hot and invigorating 105 degrees Fahrenheit. 'We bought the Indian Hot Springs twenty-five years ago when my husband was still alive. We dug out another spring scooping out the sand and rocks and the hot water just bubbled out,' says Cindy.

'Sure, I've been reading about the spaceport in the *Sentinel* [Sierra County's local paper] and I hear that this entrepreneurial guy from England is coming across. Well, good luck to him. I hope he comes and enjoys one of my baths. He'll find there's a great welcome here.'

Two blocks away, in the trim and well-used local library, the head librarian is asked for some historical background on Truth or Consequences. 'Oh, whorehouses and gambling parlours,' she says. 'But not now,' she adds hurriedly. It's now a sedate settlement with numerous retirement homes and an ageing population.

'There's isn't much for teenagers to do at night other than drive out to Elephant Butte lake and watch the sunset,' says the lady in Radio Shack, on the corner of Broadway. 'It's quiet and tranquil, and folks just like that,' she added.

The point is that tourism has been an essential element of this small town's longevity. Now it braces itself for a new kind of visitor – and also the prospect of incomers who will be working on the new port.

Hot Springs, New Mexico, depended on the visitor for generations. Then in 1950, NBC radio producer Ralph Edwards called together his staff of a top-rated game show and proposed the idea of asking America if a town or village would change its name to Truth or Consequences. The New Mexico State Tourist Board sent the news to the manager of the Hot Springs Chamber of Commerce, who could see the possibilities. Here was an opportunity to advertise the town – and differentiate it in a congested market. The town had been confused with dozens of other Hot Springs all over America – including thirty in California.

A special town referendum was held and 1,294 of the residents voted to change the town's name to Truth or Consequences, though 295 were unhappy with such frivolity. The town returned to the polls again in 1964 and again in 1967, but the vote held up. Ralph Edwards promised to return to the town every year with a festival – which he did, bringing a string of Hollywood stars as visitors. But that fame is waning; the town, with its gift shops and Indian arts and crafts, needs refreshing.

And Spaceport America will give it something extra. Yet the port is still a considerable drive out of Truth or Consequences. The tarmac road weaves its way out northeast past Elephant Butte, the picturesque lake created by the damming of the upper tributaries of the Rio Grande. Today, it is the largest expanse of recreational water in southern New Mexico. Then it is over the hills and a cross-country drive of twelve miles to

Engle. This is real Marlboro country. On a sunny Saturday afternoon, the road is utterly deserted.

Across the plain, past thousands of mesquite bushes and soap-tree yuccas and traversing the north-to-south route of the infamous *Jornada del Muerto* – the journey of the dead man – there is a sense of real solitude. The plain is named after a German trader who perished on this dry, barren route in 1670. There was no water then and it concentrates the mind now.

Today, instead of Glenn Ford it is 4 × 4 Ford. At Engle there is a level crossing on the train track which takes freight-loads of Chinese containers from El Paso to Santa Fe; three pueblo-style homes stand shaded by two pine trees and a floppy mulberry bush. This is an outpost of the Armendaris Ranch, a vast 360,000-acre estate owned by the media tycoon Ted Turner, which breeds bison, desert bighorn sheep and antelope. The ranch hand is E D Edwards, a modern-day cowboy at the wheel of his large SUV, armed with radio and global positioning to check on his valuable radio-tagged steers.

'I've heard about this spaceport – it's a few miles south. S'pose I'll believe it when I see it. This place is *sooooo* remote, I don't know how you're gonna get loads of rich customers out here,' he drawls. He says, 'Howday,' then roars his vehicle off over the rail crossing.

Turning right at Engle and heading south, parallel to the railway and the *Jornada del Muerto*, the spaceport site is another eight miles down a dusty track. Here there is only a spattering of hardy souls. One is Greg Allen, a grey, bearded former record producer and his wife Linda, a travelling nurse. He lives in a large, wooden cabin at the end of a dusty track called Abbey Road. The couple are still Beatle fans. They invented the house name – now on the map – so they could get deliveries from the Internet. Reluctantly, Greg is resigned to having to move.

'I'm not going to live here with all this disruption. That's been the attraction of this wilderness. It is so quiet and the sky at night is wonderfully clear – you can see the heavens here. I can understand why it would make a great spaceport, but I'll be moving on – perhaps back up north to the Arctic,' he sighs.

Down the road the Cains are selling up too. They are both getting on in years and Ben Cain is frail after several cancer operations. But they, and Phil and Judy Wallin, own the site for the 18,000 acre spaceport. It's the second time in three generations the Cains have been persuaded to move. The family owned land at the White Sands Missile Range before the Second World War, when it was requisitioned by the military. They moved about ten miles over the nearby San Andres mountain range to their present home. Now they will be moving too.

While the Virgin Galactic terminal and regular flights of SpaceShipTwo are some time off, Spaceport America has already been in action. New Mexico isn't the first state to dream up a commercial spaceport – Alaska's Kodiak launch complex and Mojave Airport in California already claim space launches to their credit, while the Oklahoma Spaceport has been granted a licence from the US Federal Aviation Administration.

But, on Monday 25 September 2006, a small footnote in aerospace history was written. The first rocket launched from the New Mexican spaceport. It failed to reach space, waving uncontrollably before plummeting back to Earth barely a tenth of the way into its journey. The unmanned, twenty-foot SpaceLoft XL rocket was carrying various experiments and other payloads for its planned suborbital trip seventy miles above Earth. The rocket took off at 2.14 p.m. and was supposed to drop back to Earth about thirteen minutes later at nearby White Sands Missile Range, just north of the launch site. But three miles from the launch site, witnesses saw the rocket wobble, then go into a corkscrew motion before disappearing in the clear sky.

It was a harsh reminder that rockets aren't always safe and reliable. Something radical would have to happen before thousands of sensible people would risk taking a similar type of flight. But there were significant changes afoot that would build reassurance and confidence to the new breed of space traveller.

3. A TOURIST IN ORBIT

S pace tourism has its roots in the science fiction that emerged after the terrible slaughter of the American Civil War. Jules Verne, the populist French writer, had a brilliant mind and fertile imagination, taking his readers to the centre of the Earth and 20,000 leagues under the sea. He was obsessed with the human desire to explore. But it was his fictional hero, Michel Ardan, a member of the equally fictitious Gun Club, founded during the Civil War, who trumpeted the idea of building an explosive projectile the size of the Empire State Building, in his satirical novel *From the Earth to the Moon* in 1865.

The French hero, addressing an expectant crowd in Tampa town hall in Florida, proposed his vision of going to the Moon by means of a huge cannon. 'We are going to the Moon,' exclaimed Ardan, 'We shall go to the planet, we shall go to the stars as we now go from Liverpool to New York, easily, rapidly, surely, and the atmospheric ocean will be soon crossed as the ocean of the earth! Distance is only a relative term, and will end by being reduced to zero.'

Verne's spaceship, the *Columbiad* – with an extra 'd' – carried three people, was fired into space and returned to Earth by splashing down into the sea – just like the Apollo 11 mission. While Verne's own ballistic calculations were rather batty, the idea sparked dreams of space travel that resonated for a hundred years.

In 1969, as the successful Apollo 11 capsule headed back to Earth, a primetime television transmission from the spacecraft paid tribute to Verne's vision. Each of the original Moon astronauts explained what landing on the Moon meant to them. Neil Armstrong kicked off the broadcast: 'Good evening. This is the commander of Apollo 11. A hundred years ago, Jules Verne wrote a book about a voyage to the Moon. His spaceship, *Columbia*, took off from Florida and landed in the Pacific Ocean after completing a trip to the Moon. It seems appropriate to us to share with you some of the reflections of the crew as the modern-day *Columbia* completes its rendezvous with the planet Earth and the same Pacific Ocean tomorrow.'

The Moon-mission astronauts then talked about their experiences. Mike Collins, who stayed in the command module while Neil Armstrong and Buzz Aldrin touched down on the lunar surface, praised the immense teamwork and the painstaking process of assembling a craft to go to the Moon. It was left to Aldrin to express some of the awe of the experience and reflect on the spirit of exploration – the same spirit that is driving today's generation of space tourists.

'We feel that this stands as a symbol of the insatiable curiosity of all mankind to explore the unknown. Neil's statement the other day upon first setting foot on the surface of the Moon, "This is a small step for a man but a great leap for mankind", I believe sums up these feelings very nicely.' Nearly forty years later, Buzz is still a big part in this story.

But the first real space tourists were from the Soviet Union. At 10.55 a.m. Moscow time on 12 April 1961, a small, dishevelled man wearing an orange flight suit and a white pressure helmet landed in a field watched only by a cow and a few peasants, including Anya Takhtarova. Anya was startled. Legend has it that she stepped forward gingerly and asked him: 'Have you come from outer space?'

There was a brilliant smile and the triumphant reply: 'Yes, would you believe it? I certainly have.'

Yuri Gagarin was the Earth's first space tourist – although his trip was brief and he never managed a second flight. But Gagarin is the starting point for the aspiring space tourist. While his place in history is assured as a Soviet legend, he was purely a human guinea pig fired into space to experience weightlessness and glimpse the stars. He was not much more than Jules Verne's Michel Ardan.

After the landing of his celebrated Vostok spacecraft, cosmonaut Gagarin, just 27 years old, was whisked off to the city of Saratov where he proceeded to dedicate the flight to the people of a communist society. His flight was over so quickly. It was just over an hour earlier, at 9.07 a.m. Moscow time, that the supporting arms of a giant gantry gently opened up and the huge white rocket rose into the clear blue sky. Gagarin, strapped in tightly, shouted: 'Off we go!' as he was pressed back hard into his couch. As the rocket broke through the sound barrier, there was an unbearable silence from Gagarin's capsule, and growing anxiety at ground control. Then, over the intercom, he blurted excitedly: 'I see the Earth. The loads are increasing. Feeling fine.'

The pressure reached nearly 6 Gs as the second-stage rocket boosted him out of Earth's atmosphere and into orbit. Moments later, instead of being pressed back harder in his seat, he felt himself levitating, held back by the straps as he was suspended against the couch in the tiny craft. The second-stage rocket separated, dropping back to the ground, and the world's first space tourist began sending back the first-ever postcards from space as he began reading the instruments, checking the equipment and recording the effects of weightlessness.

He broke the record for the world's fastest human cannon-ball, shooting into orbit at a speed of nearly 18,000 mph

(29,000 kph), or five miles a second – faster than a speeding bullet. While ground controllers panicked again about a brief loss of contact, Gagarin was savouring every moment of his flight, soaring over Siberia, Japan, Cape Horn and back towards Africa. He tried eating a snack in space and taking a drink. The weightlessness gave him a sense of freedom in such a confined space.

Yet he was simply a space passenger – and a trained observer. He wasn't needed to pilot the craft. The ingenious Soviet engineers had designed Vostok-1 to operate automatically from launch pad until touchdown. Along with all the emerging space travellers to come – until Virgin Galactic's revolutionary design – the first generation of astronauts and cosmonauts were basically sitting on top of a massive intercontinental ballistic missile.

The Russian cosmonauts, who were all proud and experienced pilots, didn't like their status as mere passengers. And while a manual back-up was added to Gagarin's Vostok capsule to appease this brave new breed of human beings, it was completely useless. A lock was fitted to the back-up controls with the combination numbers kept in an envelope attached to the cabin wall and well out of reach in an emergency.

Just over an hour after Gagarin's launch, the retro-rockets fired up to slow down the capsule. The Soviet cosmonaut was heading for re-entry to the Earth's atmosphere. Facing backwards, his thin frame was shaken through more than 8 Gs as the craft began its deceleration. He gazed through the porthole at a devilish fireworks display as the capsule's exterior fried to thousands of degrees and its protective coating burned away in front of his eyes. His charred capsule hit the ground and for years there was a mystery over how he survived – but the cosmonaut had managed to clamber free and parachute down to Earth.

Today, the landing site at Smelovaka, near Saratov, is a national monument, marked by a huge 40 m-high titanium obelisk, where thousands of visitors pay homage to Gagarin and the outstanding achievements of the Soviet space programme, which have come full circle. The mega-rich space tourist can now pay millions of dollars to hitch a ride on a Russian spacecraft. On 19 September 2006, Anousheh Ansari, a wealthy Iranian–American, blasted off from the steppes of Kazakhstan in a Soyuz TMA-9 spaceship with a Russian cosmonaut and an American astronaut. She was heading for a spell on the International Space Station.

Gagarin, the Soviet pioneer, was killed in a training flight on 27 March 1968, leading to widespread shock in the Soviet Union, a nation used to the disappearance and deaths of prominent figures. His ashes are buried in the Kremlin Wall alongside other heroes – and a few rogues – of the Union of Soviet Socialist Republics. Much to the chagrin of many Americans involved in the 1960s space race, Gagarin – and the Soviet Union – got there first.

Gagarin was born in the hamlet of Klushino, near Smolensk, in the western Soviet Union. The family moved to Gzhatsk – now called Gagarin – where he went to school. After leaving school he worked in a foundry before joining an aeroplane club where he gained flying experience. He prog-ressed to flying jets and then in 1960 was accepted to become a future cosmonaut. He was still a relative novice, having clocked up only 230 hours in the air.

Gagarin was sent to the newly constructed cosmonaut-training centre at Zvezdniy Gorodok, better known as Star City. He didn't get a lot of warning for his first flight, being selected only four days before his historic mission. In all, there were six Vostok missions in the cramped spherical capsule that was only 7 ft 5 in (2.3 m) in diameter.

Russia's lead was due in part to the dream of a great visionary and cosmic scientist, who came to prominence after the Soviet revolution in 1917. Konstantin Eduardovich Tsiol-kovsky, born in September 1857, spent most of his life in a log cabin on the outskirts of the Russian town of Kaluga. As a child, he caught scarlet fever and lost much of his hearing. He was uneducated yet taught himself by reading books at home, including a Russian translation of Verne's *From the Earth to the Moon*. He wasn't accepted at elementary school because of his hearing difficulties, so he stayed at home until he was sixteen. From this inauspicious start, he emerged as one of the space age's greatest visionaries.

This rocket scientist and pioneer of cosmonautics was captivated by the thought of being a tourist travelling the galaxy. He eventually found work as a high school mathematics teacher until retiring in 1920, but it was his theories on many aspects of space travel and rocket propulsion that gave the Soviet Union its technological edge. Inspired by a visit to see the Eiffel Tower in 1895, he conceived the idea of a space elevator, a device which could catapult spacecrafts up into space. In Russia, he is still considered the father of human space flight.

His most famous work, *The Exploration of Cosmic Space by Means of Reaction Devices*, was published in 1903, and is arguably the first academic treatise on rocketry, although there is an argument that claims English mathematician William Moore got there first, with a treatise on the motion of rockets and naval gunnery in 1813.

But a single equation concocted by this near-deaf recluse remains the secret of rocket science to this day. The simple equation of rocket physics was captured in equation symbols. Tsiolkovsky calculated that the escape velocity from the Earth's atmosphere into orbit – known as Delta V – was 8 km per second and that, to achieve this, a multi-stage rocket fuelled by liquid oxygen and liquid hydrogen would be necessary. This calculation has been refined to 11.3 km per second – and for an object to stay in orbit at 150 miles (242 km) above the Earth, it needs to travel at 17,000 miles per hour, slightly less than escape velocity. But his calculations were an exceptional piece of work.

The definition of such a machine is simple. A rocket is a device that can accelerate by expelling part of its mass at high speed in the opposite direction of its forward movement.

While Tsiolkovsky laid the theoretical foundations for a method to break the bonds of Earth's gravity and reach for space, it was his log-cabin dream of flying to other planets in our solar system that is likely to resonate most with today's space tourist. He was the kind of dreamer who would have loved to have been able to take a trip.

'All the universe is full of the life of perfect creatures,' he said in his book *The Scientific Ethics* in 1930. He was literally a starstruck dreamer. Tsiolkovsky was as interested in the philosophy of space travel as he was with the engineering prowess needed to make space flight possible. In 1932 he wrote *The Cosmic Philosophy*, saying there should be happiness not only for all of humanity, but for all living beings in the universe. He believed that human occupation of space was inevitable and would drive human evolution. In an atheistic Soviet state, where logical positivism was the predominant credo, he defines happiness as the absence of suffering throughout the universe, for all times.

This ideal of a Cosmic Utopia might well have been scripted by Gene Roddenberry, the creator of *Star Trek*. It certainly conjures images of Captain Kirk and the interplanetary missions of the Starship *Enterprise*. But the Russian was serious, arguing that humankind's overriding task was to study the laws that rule the stars. 'To do so, we must study the universe, and therefore we must learn how to live in outer space. To begin that long period of our evolution, we will have to design large manned space rockets.'

He believed it was human destiny to occupy the solar system and then to expand into the depth of the cosmos, living off the energy of the stars to create a cosmic civilisation that would master nature, abolish natural catastrophes, and achieve happiness for all. This was radical thought in 1926 when Tsiolkovsky defined his Plan of Space Exploration, consisting of a shopping list of steps for human expansion into space.

So how did he fare? Pretty well indeed. Will Whitehorn, Virgin Galactic's president, would have relished a vodka-fuelled evening with this great man.

Tsiolkovsky imagined the creation of rocket airplanes with wings. He talked of the experimental use of plants to make an artificial atmosphere in spaceships, and of orbiting greenhouses filled with plants. In further thoughts, he expressed his vision of large orbital habitats around the Earth, using solar radiation to grow food, heat the astronauts' quarters, and fuel transport throughout the solar system. He believed that the entire solar system and beyond would one day be colonised and, as our sun begins to die, that human beings remaining would go off to find other suns.

Such powerful ideas inspired humans throughout the twentieth century and into the third millennium. Tsiolkovsky's work was given powerful impetus with the rapid development of rocket and space technology from the end of the Second World War onwards.

The adopted son of the American space rocket was Wernher von Braun. His career had a remarkable upwards trajectory that would have confounded the physics of rocket science. He was one of the German scientists who, following the collapse of Hitler's Third Reich, was enlisted into America's secretive Overcast mission. The Americans chose to overlook the fact that the scientists had been members of the Nazi Party and should have been tried for war crimes. Indeed, von Braun's deadly V2 rocket began to terrorise London and the southeast of England in 1944, when some 3,600 were fired into Britain.

With his Nazi past suppressed, von Braun became one of the most important figures in the US space programme. Born on 23 March 1912 in Wirsitz, now in Poland, von Braun was interested in rockets from an early age. He was a wunderkind who enjoyed making toys go bang, and this led to him studying engineering at Berlin and Zurich universities. In 1930, the eighteen-year-old read an article on travelling to the Moon and it fired his imagination for space travel. It was the start of his urge to soar through the heavens and explore the mysteries of the universe.

For more than a decade, the German military had been fully aware of the potential for this destructive science and state-of-the-art weaponry. The fledgling rocket programme was transferred to the Nazi war machine and von Braun, still only twenty, was made chief of the Kummersdorf station. Hitler could see that the only way to defeat Great Britain's Royal

Navy in a future war and a strangulating blockade was to develop German rockets. By 1938, von Braun's team and his scientists, now working at Peenemunde, developed a prototype of the V2 missile. It was capable of carrying a tonne of explosives nearly 200 miles. Coupled with the Nazi's development of the bi-fuel rocket that propelled the Messerschmitt Me163B *Komet* at speeds of up to 593 mph (burning immense quantities of hydrogen peroxide and a catalyst of hydrazine hydrate in methanol), it might well have won the war for Hitler.

In an interview with the *New Yorker*, he expressed a 'genuine regret that our missile, born of idealism ... had joined in the business of killing. We had designed it to blaze the trail to other planets, not to destroy our own.' But one German colleague recalls that von Braun and his team toasted their bombing success on London with champagne.

Von Braun's war was coming to an end, though. He was shrewd enough to avoid the advancing Russian army and hide from the SS before surrendering to American troops. At first the American intelligence officers thought he was too young to be such a celebrated scientist but they soon realised they had a genius on their hands.

He was influenced by the pioneering work of Robert Goddard, the granddaddy of American rocketry.

The US press popularised some of Goddard's writing about how rockets could get to the Moon and sarcastically dubbed him 'Moon Man'. The American Interplanetary Society, a professional organisation, was set up in New York City in 1930 but changed its name to the American Rocket Society because the whole idea of interplanetary travel was being ridiculed. But the American Rocket Society became the forerunner of the American Institute of Aeronautics and Astronautics, the principal technical society in the US devoted to aerospace. The Guggenheim Foundation funded some of Goddard's research, but as the Second World War approached the US Army and Navy remained uninterested in rockets. This was an arrogant oversight. Indeed, one of Goddard's proposals for support to the Air Corps met with rejection. His letter said the corps 'was deeply interested in the research work being carried out ... under the auspices of the Guggenheim Foundation but does not, at this time, feel justified in obligating further funds for basic jet propulsion research and experimentation'.

Pearl Harbor changed all that. By 1945, the US government rocket programme was kick-started by President Roosevelt's National Defense Research Committee. There were now large budgets and brand-new production facilities co-ordinated across the military services. Von Braun was a key hiring.

Operation Overcast – later called Operation Paperclip – brought together some of the finest scientists of the Third Reich and gave them a safe passage to a new life in America. Project Paperclip was meant to prohibit the use of war criminals and those with Nazi Party links, and the initial security report on von Braun in 1947 indicated that he was an SS officer. While there was no firm proof he was classed as a potential security risk, but this report was suppressed. A new report was issued suggesting von Braun's Nazi Party membership was simply expediency and not a security threat to the US.

This whitewashed report paved the way for von Braun to remain in the United States and obtain US citizenship. But, more importantly, it allowed the US to use von Braun's brain in its space race with the Soviet Union – and the emerging Cold War. Von Braun was a star prize for America and became the linchpin of the USA's space programme. His rehabilitation took him to Huntsville, Alabama, in the 1950s and he was instrumental in the development of the Redstone ballistic missile and the Jupiter intermediate-range ballistic missile, both early stages of the Explorer, the USA's first satellite launch, and the Pioneer satellite.

In 1960, von Braun became the Director of NASA's George C Marshall flight centre, which developed all the Saturn booster rockets that sent the Apollo capsules into space. One of his supreme achievements was the Mercury programme, which ran from 1959 to 1963, with the goal to put a man in orbit around the Earth. John Glenn became the all-American hero on 20 February 1962 when Friendship 7 orbited the Earth, but it was Little Joe, Redstone and Atlas, von Braun's three booster rockets, that kept the programme on track. He maintained a 100-per-cent record of successful launches while the world's largest rocket, the Saturn V, was being developed.

His celebrity extended beyond the sphere of rocket building. He wrote several books and was unable to resist the lure of Hollywood, where he became the technical adviser on several Walt Disney films about space flight. He was even the inspiration for Professor Ludwig von Drake, the Disney cartoon character who looked like a duck but spoke with a German accent. He died on 16 June 1977 in Alexandria, Virginia, having seen his dreams realised through the Apollo programme.

Von Braun was the rocket supremo for the lunar missions. Neil Armstrong, Mike Collins and Buzz Aldrin had great faith in the Saturn rocket, but no one could ever be completely sure about any rocket's performance, not even the massive machine designed by Von Braun's team of rocketeers at the Marshall Space Flight Center, with its 12 million working parts. Even von Braun had moments of doubt: 'When we wheel out one of the rockets to the launch pad, I find myself thinking of the thousands of parts – and all built by the lowest bidder – and I pray that everyone has done his homework.'

'It was certainly a very high-performance machine,' Neil Armstrong told his biographer James Hansen. 'It was not perfect, though. Indeed, in the flight after ours, there would be problems.' A lightning strike just after Apollo 12's takeoff temporarily knocked out the electrical systems on the spacecraft. This failure was not the rocket's fault except that an ionised plume from the rocket made the entire vehicle more electromagnetically attractive.

In the late 1950s, Von Braun's nemesis was the Soviet Union's chief space designer, Sergei Korolev. He had laid out his plans for a manned mission to Mars, which has the environment closest to Earth's in our solar system. The Aelita project – known as the Heavy Interplanetary Manned Vehicle project – was started in the Soviet Union at the OKB-1 bureau, now RSC Energia, and would involve putting a spacecraft with a starting mass of 1,630 tonnes into orbit. This would be assembled in space before heading to Mars.

In 1959, a team of Soviet enthusiasts was already working on a concept of circular orbit for a spacecraft that would then be hurled into a Mars fly-by trajectory. Then, assisted by the gravity field of Mars, it was to come back to the vicinity of the Earth, with the descent vehicle returning to Earth. In this flimsy pressurised craft, three space tourists would live for two to three years in a confined cabin six metres in diameter.

The psychology of such a human experiment has been the source of countless masters theses, suggesting the confined space travellers might go insane before reaching their destination. The cockpit would double as a radiation shelter for the crew during solar-flare activity, while the vehicle would revolve on its axis to create artificial gravity.

For the Russians, the first step was to keep putting humans into space – and they were brutal in their determination to succeed. Four months after Gagarin, Vostok-2 took the second space tourist into orbit. Gherman Titov was a clean-cut handsome Soviet hero, aged only 25. Born in Siberia, he was enrolled at an elementary aviation school at thirteen and nine years later came top of his class as an air force pilot.

Titov was more a medical guinea pig than a tourist. Korolev said he wanted to find out what impact space flight had on a man's organism, so Titov was whisked into orbit for a flight lasting over 25 hours and involving 17 orbits of the Earth. He felt well enough after the first orbit, but then succumbed to space sickness – nausea and dizziness and an inner ear problem that began to give him difficulty with his balance and motion in the weightless environment.

Titov was the first man to sleep in space – which certainly helped his sickness. After eight hours, he had a meal using special tubes. Then, after seventeen trips around the planet, the Vostok automatically fired its retro-rockets and headed back to Earth.

Tensions in the world had been increasing since the Russian launch of Sputnik 1 on 4 October 1957. The Americans were freaking out at the success of the Soviet space programme.

As Tom Wolfe puts it in his book, *The Right Stuff*: 'A colossal panic was under way, with Congressmen and newspapers leading a huge pack that was baying at the sky where the hundred-pound Soviet satellite kept beeping around the world.'

Sputnik was a momentous event in the developing Cold War, and a serious escalation after the Soviet development of the atomic bomb in 1953. It meant the red menace now had the ability of delivering the bomb – using an intercontinental ballistic missile – right into the heart of the American homeland.

'The Soviet programme gave off an aura of sorcery,' wrote Wolfe. 'The Soviets released practically no figures, pictures or diagrams. And no names: it was revealed only that the Soviet programme was guided by a mysterious individual known as "the Chief Designer". But his powers were indisputable! Every time the United States announced a great space experiment, the Chief Designer accomplished it first, in the most startling fashion.'

The Chief Designer was, of course, Korolev, who was only identified in the West after his untimely death in January 1966 as a result of a haemorrhoid operation that went badly wrong. By 1962, the Soviet Union was ready to deliver another dose of propaganda to show that it was technically superior to the United States. President Nikita Khrushchev sanctioned a real show with the next two Vostok launches. By now, the dictator enjoyed mocking the democracy-hugging Americans. The idea was to launch two craft – Vostok-3 and Vostok-4 – one after each other to put themselves into close orbit for future docking manoeuvres. This would mean a longer stay in space.

So, a year after Titov's space sickness, Vostok-3 was launched with 32-year-old Major Andrian Nikolaev inside. More firsts were recorded by the Soviet Union, with live black-and-white television pictures beamed to an excited Soviet nation in desperate need of some cheering up. The Russians created a further coup de théâtre when the next day they launched Vostok-4. The fourth Soviet man in space was a 31-year-old air force lieutenant colonel, Pavel Popovich, who was blasted into orbit. Nik and Pop – as the popular media dubbed them – passed close to each other in space. It was a spectacular achievement, like a huge Russian circus finale. The American media talked of the widening of the 'space gap'. Yet, just like their predecessors, these men were nothing more than onlookers.

4. AMERICA PLAYS CATCH-UP

T he Americans had egg on their faces. They were forced to
play catch-up in space. President John F Kennedy made a
promise to his nation. In a momentous speech at Rice
University in Houston, Texas, on 12 September 1962, the
charismatic president talked about the challenges facing
America. He said the country was built not by those who
waited and rested but by those willing to take America
forward – and this meant the exploration of space. 'We need
to be a part of it and we need to lead it,' he said.

Space had to be explored and mastered. 'We choose to go
to the Moon. We choose to go to the Moon in this decade and
do the other things not because they are easy but because they
are hard ... because that challenge is one we are willing to
accept.'

Ironically, this race to the Moon signalled the death knell
of an ambitious programme that might have developed a
separate spaceship controlled by the pilot. Alan Shepard had
been able to restore some of America's wounded pride in May

1961 when, within weeks of Yuri Gagarin's triumphant flight as the first space tourist, the Stars and Stripes were up in space too. Shepard's first manned Mercury spacecraft reached an altitude of 116 miles and travelled at a speed of 5,180 mph. Surely no plane could match that? Yet an experimental project was giving a good account of itself.

Just over two years after Shepard's first flight, on 22 August 1963, with the Beatles at the top of the Billboard charts, NACA (the National Advisory Committee for Aeronautics) test pilot Joseph 'Joe' Walker reached an altitude of 354,200 ft – 67 miles high – in a rocket-powered X-15 jet, controlled by Walker himself. It returned to the Edwards Air Force Base runway from where it had taken off. It had been attached to the underside of a converted B-52 mother ship and the vehicle was used again and again.

The X-15 was an experimental hybrid, not really a spacecraft but a lot more than an aeroplane. The boffins of NACA – the predecessor to NASA, which was formed in 1958 – came up with the idea in the 1950s and it was built by North American Aviation – later known as the Rockwell Corporation. This incredible machine was the fastest and highest-flying winged vehicle ever built. It was designed to explore the lower end of hypersonic flight, from Mach 5 to Mach 6 (over 4,000 mph), and reach an altitude of 200,000 ft – or 38 miles high. It was first flown in June 1959. Nearly fifty years later, Virgin Galactic's SpaceShipTwo will surpass the achievements of the outstanding X-15 programme from the early 1960s, but there are many similarities.

The maximum speed achieved by the X-15 was 4,534 mph (about Mach 6.7) – almost the speed of the Redstone rocket that fired Shepard, and the other 'magnificent' seven Mercury astronauts, including Gus Grissom and the US's first orbital tourist, John Glenn, into space.

Burt Rutan, the builder of SpaceShipOne and now SpaceshipTwo, reveres the X-15 programme as one of the best aerospace projects ever devised. A wealth of information was garnered during this X-15 era, with Neil Armstrong, the first man on the Moon, himself contributing specific papers on supersonic flight-test instrumentation. Years later, the declass-ified technical files were pulled out of the cupboard, dusted down and pored over again by Rutan at his Mojave office.

So how much was the SpaceShipOne project influenced by the costly X-15 programme? Brian Binnie, Virgin Galactic's test pilot, explains: 'The X-15 programme was primarily interested in characterising the heating effects of high-speed flight and to better understand the re-entry aerodynamics at various dynamic pressures,' he said. 'Midway through the programme, though, they set their sights on making some high-altitude flights using a profile which more closely resem-bled that flown by SpaceShipOne. Nonetheless, it is note-worthy that SS1 is about half the size, weights a fifth as much, carried three times the payload and had better performance. Costs are a little harder to compare but we spend less than 5 per cent of the X-15 budget at the time they went for the altitude flights. Of course, having a luminary like Burt Rutan at the helm will beat any bureaucracy any day.'

Plus the creators of SpaceShipOne had the advantage of new, lightweight, composite material and computing power way beyond the imagination of X-15's original engineer Harrison Storms and his band of slide-rule slaves. But more of this later.

'The X-15 needed several hundred miles of space to fly the hypersonic trajectories it would fly,' said Neil Armstrong. 'I was involved in the development of this high-speed range, or High Range, and the combination of radar, communications and telemetry that would be required to get data quickly, accurately and in a minimum amount of time. The airplanes were a big investment, and the cost per flight was high, so it was important to be able to maximise the efficiency of getting the data.'

Virgin Galactic's space tourists will follow the trail of those early flights by Armstrong, who reached out for space during his time as an X-15 test pilot. Today's space tourist will also be able to appreciate Armstrong's feats in the X-15's forerun-ner, the F-104, as captured in *First Man*, James Hansen's biography of Armstrong.

'Streaking upward past 45,000 ft, he [Armstrong] passed the biological threshold at which a person could survive without the protection of a spacesuit. When his near-vertical climb reached 90,000 ft, atmospheric pressure fell to a scant six millibars, about 1 per cent of the pressure of sea level. Outside the cockpit, the temperature dipped to minus-sixty degrees Fahrenheit.'

Thankfully, SpaceShipTwo will not be emulating Neil Armstrong's next manoeuvre, but what he achieved was well noted by Rutan, Virgin Galactic's chief technology officer, George Whittinghill, and the Scaled Composites spacecraft designers.

'The only way to control his plane at the top of its ballistic arc was to invoke Newton's Third Law and expel some steam via jets of hydrogen peroxide. A pilot in a near vacuum could manoeuvre his airplane in pitch, yaw and roll just as manned spacecraft would later do.'

Newton's Third Law of Physics is that for every action, there is an equal and opposite reaction. Or, as the old adage states: what goes up, must come down. Now, with all the energy from Armstrong's F-104 dissipating, the jet came close to a virtual standstill on its tail. At this stage, the main engine was switched off and hydrogen peroxide rockets in the wingtips were used to boost the rocket higher.

'For over half a minute at the top of his climb, he experienced the feeling of weightlessness,' writes Hansen. This sensation will certainly be shared by the Virgin Galactic pioneers, but be assured Armstrong's return to terra firma will not be simulated. 'Streaking down nose-first into the atmosphere, enough air molecules eventually passed through the jet's intake ducts to allow Armstrong to restart the engine, and at a speed of about Mach 1.8, begin his recovery from the unpowered dive.'

Armstrong soon moved on to the X-15, flying it seven times before joining the elite group of NASA individuals who claimed to be real astronauts. On 20 April 1962, in his sixth X-15 flight, he reached 207,500 ft, just over 40 miles high. Was he in space yet? Not quite.

The rivalry between NASA and the US military threw up an intriguing question. Where does space actually start? The military believed that a flight above 50 miles high qualified the pilot as an astronaut, but NASA disagreed. So did the Fédération Aéronautique Internationale, custodians of the world's aviation records. It said it started at 62 miles, or 328,099 ft. The emerging generation of space tourists will be going beyond this and so officially qualify as astronauts.

Joe Walker became NASA's prime pilot after Iven Kincheloe, a rugged combat hero, was killed in a routine flight in an F-104. A red warning light on his control panel lit up, giving him a split second to make a life-or-death decision. His ejector seat fired him out sideways; it was the wrong choice. These were the pioneering days of the Mercury space project and test pilots Walker, Neil Armstrong, Crossfield and White were far keener on flying the X-15 than becoming 'Spam in a can', the Edwards Air Force Base nickname for Mercury. As Tom Wolfe puts it in *The Right Stuff*: 'The testing of the X-15 would proceed, in order to develop a true spacecraft, a ship that a pilot could fly into space and fly back again down through the atmosphere for a landing. Much was made of the fact that the X-15 would "land with dignity" rather than splash down in the water like the proposed Mercury capsule.'

But it was only in August 2005 that NASA finally recognised the achievements of Joe Walker, awarding him his astronaut's wings posthumously.

Armstrong's X-15 was powered by the XLR-11, an experimental liquid rocket engine built by the Reaction Motors Division of Thoikol Chemical Corporation. It was a powerful enough beast. Reaction Motors (RMI) of Pompton Plains, New Jersey, was already famed for making the 6000C-4 power plant – the XLR-11 used in the Bell X-1 to break the sound barrier, a speed of 761 miles an hour at sea level and 660 miles an hour at 40,000 ft.

It was the legendary Charles E Yeager who broke this record on 14 October 1947 in his bright-orange stubby rocket plane, now hanging in the Smithsonian Institute's Air and Space Museum in Washington alongside Rutan's Space-

ShipOne. The X-1 was loaded in the modified bomb bay of a Boeing B-29 Superfortress, filled with a dangerous cocktail of liquid oxygen, alcohol and nitrogen, which was used to push the oxidiser and fuel into the rocket engine. The whole vehicle was a lethal flying bomb. As dawn broke over the Muroc Dry Lake, the US Air Force's test-flight centre [now the Edwards Air Force Base] in the Mojave Desert, Captain 'Chuck' Yeager, nursing two broken ribs from a fall off his horse, scrambled into the X-1 named *Glamorous Glennis* after his wife. It was a primitive era, compared to today's digital processing and pressurised environment. Yeager wore a home-made helmet, attached his oxygen mask, and plugged in his microphone and headset. He donned a light leather jacket over his flight suit, although the liquid oxygen was numbingly cold.

The B-29 took off on its historic flight, with Major Robert Cardenas flying the Second World War bomber. Yeager began pressurising the fuel tank and disconnecting the nitrogen hose, before Cardenas put his plane into a shallow dive to gain speed before releasing the X-1 into the sky. As the four 6000C-4 rocket chambers – made of seamless stainless steel from Pittsburgh – fired off, Yeager began a bone-shaking and terrifying ride. As the X-1 approached Mach 1, Yeager grimly watching the Machmeter edge towards the record, the engineers on the ground, including chief designer Jack Ridley, knew that Yeager had broken the barrier. They heard the double explosion of a sonic boom for the first time. The Bell flew at Mach 1.06 – seven hundred miles per hour.

Apart from breaking the record, the X-1 made another significant contribution to US military strength. During the Korean War between 1950 and 1953, the US Air Force faced the advanced Soviet-designed MiG-15 jet, which was faster and easier to manoeuvre than its American counterpart, the F-86 Sabre. The MiG was a better plane in every department, except one. The Sabre had a much more stable gun platform because of an adjustable elevator developed on the X-1. This gave the US Air Force a ten-to-one victory ratio over the MiG-15s of the Chinese, North Korean and Soviet air forces.

The X-1's success led to the continuing use and development of the XLR-11, which was the forerunner to the more powerful XLR-99, the first restartable liquid propellant rocket engine. This was another step change. It allowed a pilot to pull back the throttle and control the thrust. It gave the pilot real control, allowing him to shut down the engine on the edge of space, and be able to start it up again on the descent. It delivered 57,000 lb of thrust and was propelled by a combination of liquid oxygen and anhydrous ammonia, fed into the engine by pumps powered by hydrogen peroxide – the substance used to dye so many of the world's women blonde. The XLR-11 comprised two rocket motors; each motor had four chambers and each chamber delivered 1,500 lb of thrust, or 12,000 lb in total.

Armstrong never flew the more powerful machine with the XLR-99 engine, delivering five times more power. Test pilot Scott Crossfield, who was the first person to fly Mach 2 in a Douglas D-588-2 Skyrocket, was at the throttle for the first powered XLR-11 flight on 15 November 1960 when it was launched from a B-52A.

Virgin Galactic's George Whittinghill, who worked at NASA and is a rocket propulsion expert, says: 'The X-15's trajectory was ballistic. It would just go up and come down like a shell or a bullet. It didn't have the energy to go orbital.'

Returning to Earth safely remained an insoluble challenge – the Holy Grail of space flight. And the second phase of the X-15 programme was intended to show how a pilot could re-enter the Earth's gravity from orbital velocity. The Dyna-Soar – known as the X-20 – was designed to prove a pilot could control such a daring and dangerous descent. Burt Rutan has cracked this with his revolutionary re-entry mechanism on SpaceShipTwo. This is a unique safety feature – an exotic feathering wing allowing the space vehicle to float down to Earth in a controlled descent like a sycamore leaf. Such innovation is possible only with the advances of composite carbon materials. Back in the 1960s, the American project fell victim to intense political bickering and a functional X-20 was never built, though many of its systems were incorporated in other experimental research. Burt picked up a great deal on the cheap from those expensive lessons of the past.

What is certain is that the SpaceShipTwo and several other of the new generation of suborbital vehicles will be dropped by a mother ship before rocketing into space. And a pilot will be at the controls. Just like the X-15.

5. A BEAUTIFUL BLOGGER BLASTS OFF

The competition to take paying passengers is hotting up. But Eric Anderson, the president and chief executive of Space Adventures, is supremely confident. He has already arranged for some ultra-rich passengers to go into orbit. He buzzes around with the energy of a charged electron. And in this contest to win paying passengers, there is a delightful *frisson* between Anderson and his Virgin Galactic rivals. Anderson wants to be the first. In fact, he is desperate to get there first. And in the opening hands of the poker game of space, he has played a few winning aces.

The NextFest, run by *Wired* magazine in New York City, is a mammoth hot tub for new-technology buffs. While the buzz of noisy schoolchildren could be heard outside during education day, around 500 people were listening intently to Anderson's presentation in the cavernous Javits conference centre.

'I am unbelievably excited to talk to you all about this today. How many of the ladies in the audience would like to go into space in their lifetime?' he asked. A forest of female hands shot up. Bespectacled Anderson, standing in front of a giant screen with an infrared fob in his right hand, looked mighty pleased.

'Several,' he said. 'And there's lots of guys out there too! We just launched the first woman as a private paying passenger. Her name is . . .'

Before he could finish, and flick the fob button, he was interrupted by a spontaneous outburst of applause that rang out around the hall.

'Very good. Yes, yes. Anousheh Ansari. She just came back to Earth on Thursday and I'm going to show you some video.'

More applause.

'My company is Space Adventures and we've been around for nearly ten years. We are the company that takes people into orbit. We are also working on suborbital space flight too. We have a partnership, with the venture capital group that is owned by the woman who has just flown into orbit, to build suborbital space vehicles, which will be flying from spaceports in Singapore, in Dubai and also in the US. But we've not yet announced where that will be.

'Let me tell you a little more about it.'

Anderson only had time to punch out the bullet points of his business, but it was an impressive start. For sure, his company plans to launch orbital flights from the United Arab Emirates' Ras Al-Khaimah airport and from Changi airport in Singapore, but he didn't have time to explain that his Explorer vehicle is being designed by Myasishchev, using an air-launched C-21 rocket. A Russian design bureau was founded in 1951 by Vladimir Myasishchev, one of the Soviet Union's most illustrious aerospace engineers, and his legacy and the evolution of his thinking lives on today.

He continued: 'So far there have been four space tourists. Four who have paid their way to go into space. The first was Dennis Tito, the second Mark Shuttleworth, the third was Greg Olsen and, more recently, and the one you've heard about the most is . . .'

Anderson aimed his fob and pressed again. Anousheh Ansari's beautiful – and pixellated – face immediately appeared on a giant screen behind him. Her voice boomed out across the sound system. 'I certainly hope that space travel will become something that will be acceptable to everyone who wants to take the trip. I will personally do everything in my power to see that happen,' she said.

Her flight was a public relations executive's dream. She was a slice of glamour in space with her thick curly hair and her dark olive-skinned Persian looks. She was articulate and committed. Originally she had been the back-up astronaut, but the Japanese entrepreneur, Daisuke Enomoto, known as Dice-K, was forced to pull out. He would have been the first Japanese commercial tourist in space, but on 21 August 2006, the Russians announced that Dice-K was 'deemed not ready to fly for exclusively medical reasons'.

Less than four weeks later, Anousheh's dream would be a reality. On 18 September 2006, she blasted off on her ten-day expedition to the International Space Station. As a successful entrepreneur, and co-founder of Prodea Systems, she had become so wealthy the $10 m price tag was pocket money – her company, Telecom Technologies Inc., was bought by Sonus Networks in 2000 for $750 m.

'It is so rewarding that she has been able to go into orbit in the last few weeks and make her dream come true and inspire millions of people out there, especially women. It is something that they can do,' said Anderson, as he narrated the next sequence of video.

Anderson outlined how Anousheh spent several months training at Star City, learning Russian, and undertaking drills and testing before getting the go-ahead to fly on a Russian Soyuz spacecraft to the International Space Station, where she would circle the Earth once every ninety minutes.

'She called me from her satellite phone right after landing to tell me that it is one of her life goals to communicate that experience and share it with all the millions of people out there who want to go into space,' he told the crowd.

Anderson flicked his fob once more and a colour video from the International Space Station shows Ansari cross-legged in space juggling with an apple in zero gravity. 'Weightlessness is absolutely fantastic. It is impossible to describe, but it is like floating like a feather in the air. Effortless. You don't need to try. It's a wonderful feeling after you get over the motion sickness, of course,' she said on the film.

Her ten days in orbit included two spent in a tiny Soyuz craft chasing the orbit of the ISS. Just before opening the docking hatch her Russian colleague asked her to savour the fragrance of space, though Ansari described it as a smell more reminiscent of a cheap chemical air freshener. But the startling views out towards the cosmos stunned her.

'I have never seen so many stars and it is all clear. The skies and the Milky Way is just . . . I can't describe, it is so amazing. Anyone who would like to see, it's . . .' Her words faded away.

What was immensely gratifying for Ansari was the world-wide response. Every day in orbit she filed her blog – a personal diary of her day – on the Internet. It generated six million clicks and ten thousand emails a day. Thousands of people, inspired by her warmth and open demeanour, took to this woman who was enjoying her fifteen minutes of orbital fame.

So Space Adventures is trying to get ahead of the pack. It plans to use a Russian-built Explorer vehicle, designed by the Myasishchev bureau for $10 m, to carry five people into suborbit. 'We will not disclose the development schedule until it is finalised. We have the utmost confidence we will enable operations of the world's first commercial suborbital flights,' said Anderson.

So far, Space Adventures has said the Explorer *Cosmopolis XXI* – known as the C-21 – will ride on the back of an M-55 Geophysika high-altitude research aircraft. This aircraft was originally flown to shoot down weather balloons that were flown over the Soviet Union by the CIA. The Russians derived the design of the aircraft from the wreckage of the U2, the US spy plane flown by Gary Powers that was shot down over the Soviet Union in 1960. The C-21 is a three-seater suborbital reusable launch vehicle. Now the same team who designed the

Russian space shuttle *Buran*, which flew only once on 15 November 1988 and was shelved to save costs, are working on this project. After taking off, the M-55 carrier climbs to twelve miles high, then the Explorer – carried on top of the plane – detaches and fires its rocket boosters, taking it into space. It can reach an altitude of 62 miles, before performing a ballistic descent with the use of a parachute to help it glide back to the ground.

So how did space tourism reach this heady stage?

Back in May 2006, at a conference of space tourism at the Royal Aeronautical Society in London, Jane Reifert, an American businesswoman, gave an insight into how space was fast becoming a commercial proposition. Reifert explained that her company, Incredible Adventures, is also planning to take three passengers into orbit with Rocketplane XP, which will take off from the Oklahoma Spaceport like a conventional airplane.

'Incredible Adventures started by accident. In 1993, Kent Ertugrul, an American entrepreneur, was in Moscow looking for some computer programmers to work for his Florida-based voice-mail company. While there, he just happened to meet the neighbour of a Russian test pilot, who offered to arrange a flight for him in a MiG29 – for the right price.' Ertugrul loved his stomach-wrenching MiG flight so much that he obtained the rights to market the flights worldwide. He returned home to Sarasota and set up the business called MIGS etc.,' said Reifert.

The business didn't set out to become a space tourism company, it just happened by accident. A chance meeting in Moscow and an advert in the *Wall Street Journal* was all it took to launch the 'fighter-jet tourism' industry. Customers began lining up to place deposits on MiG flights, much like customers are lining up today to place deposits on suborbital flights.

Word spread quickly that Americans and rich Europeans and Japanese were paying significant sums of money for jet-fighter flights. Representatives of the Russian Space Agency contacted Reifert's company and asked if they would be interested in selling space training. It was too good an offer to turn down. The company's first zero-gravity flight took place in February 1994, planting the seeds of a new space-tourism industry. The historic flight of what became known as the Vomit Comet was covered by the CNN and CBS networks and major newspapers around the globe.

In 1996, under new ownership and the new name Incredible Adventures, the company worked to get one of its customers on board a Soyuz flight to the Mir Space Station. A deposit was paid to the Russian Space Agency, but the individual failed to secure the funds needed to complete his journey, missing his opportunity to become the first space tourist.

Over a decade later, Incredible Adventures continues to offer MiG flights over Moscow and cosmonaut training at Star City. The company has seen interest in its space-related offerings grow considerably.

'In 1993, no one wanted to fly the MiG-25 and today the jet, known for its ability to fly eighty thousand feet high to the edge of space, is our most popular product. Now, in addition to Russian space trips, Incredible Adventures markets zero-gravity adventures in Florida on behalf of Zero G Corp and suborbital flights on behalf of Oklahoma-based Rocketplane XP,' Reifert said.

What Incredible Adventures is offering soon is a trip into space in a converted Lear jet powered by two 3,000 lb turbojets and then a single 30,000 lb thrust rocket engine to get into suborbit. The patented ORBITEC system – Vortex Combustions Cold-Wall Chamber engine – is designed to keep costs and weight down and increase reliability and performance. Liquid oxygen and kerosene are the main fuels – although the whole trip is like being in a plane. The XP won't be launched from a mother ship, but will take off and land from a spaceport runway. The paying passengers will reach space and have around four minutes of weightlessness, but there won't be much room for floating around.

Christopher Faranetta picked up the story to explain the evolution from the MiG adventures and Incredible Adventures to the first paying passengers. He is vice president of the Orbital Spaceflight Programme for Space Adventures. He helped forge early joint American and Russian space efforts while serving as international liaison to the Space Studies Institute, founded by the late Dr Gerard O'Neill, professor of physics at Princeton University, in 1977. He was the visionary who foresaw a modern society powered by clean, dependable, full-time solar energy supplied by solar-powered satellites. Faranetta moved on to establish and co-manage the American office for Rocket Space Corporation Energia, the largest Russian aerospace company and the prime contractor for the Russian segment of the ISS, spending ten years as deputy managing director of American operations for RSC Energia.

'When I was working for Rocket Space Corporation, our primary customer was NASA. Then one day I received a call. On the other end of the phone was a young guy called Eric Anderson. He claimed he had a customer who wanted to fly to the Mir space station,' recalled Faranetta.

Chris Faranetta and his partner Jeff Banbury put him on speakerphone and started laughing quietly with disbelief. They almost wanted to hang up the phone because they thought he was a nut. Chris's primary concern was that NASA wouldn't like it, and wouldn't be happy.

'Eric is now my boss at Space Adventures,' he said. 'Now private space exploration is good for the industry – it's good for NASA. It's good for everybody. We are essentially the only operational private space exploration company.'

Founded in 1997, with offices in Washington DC, Moscow, Tokyo and Kennedy Space Center, Space Adventures is able to use its clout and connections. 'We get a lot of input from former astronauts and cosmonauts,' says Faranetta. 'They are one of the single greatest assets of our company. They keep us on the straight and narrow and help us make the right decisions.'

Space Adventures has practical experience with the Soyuz booster and the TMA spacecraft, which is Russia's most up-to-date technology. 'It's a very simple and reliable system. It has a specific mission to deliver a crew to the International Space Station. One of the problems with the Space Shuttle is that it tries to be everything to everyone and that adds to its complexity.

'We have a space-travel experience with a destination – we fly to the International Space Station which has interior space equivalent to a Boeing 757 aircraft. As it gets bigger we will have more space. It is there and it is in orbit.'

The International Space Station needs friends. The 420-tonne structure is a modern marvel now orbiting around 250 miles (400 km) above the Earth. Over 100,000 people have been involved in its construction, with the United States, Russia, Japan, Canada, Brazil and 16 countries from Europe involved in this colossal $100 bn project. It is a symbol of how national interests can be put aside for a loftier goal. Yet every day the clock is ticking. As the ISS gets older, more time is spent fixing and maintaining it. It has burst its vast budget and is costing a huge amount to keep in space. So there are increasing opportunities to take paying passengers up into space to offset the work that still needs to be done.

'One of the great benefits of our programme is that we are asking people to go into space to do something useful. It is a requirement that anyone going to the ISS does some science when they are up there. It's a necessary part of the contract,' said Faranetta.

'Our first client was Dennis Tito in 2001 and he started out as an engineer in the JPL laboratory. He was an orbital dynamicist and he developed this program at JPL that was used to help calculate the orbits of the Mariner spacecraft mission to Mars.'

Tito wasn't a typical space tourist. Born in August 1940, he had a profound interest in science and grew up in a tough neighbourhood of Queens in New York, the son of Italian immigrant parents who 'truly believed in the American dream'. He gained a Bachelor of Science degree from New York University College of Engineering and a Masters at the Rensselaer Polytechnic Institute, before later taking a PhD in finance at UCLA's Anderson School of Business. He joined the Jet Propulsion Laboratory, near Pasadena, at 23.

JPL has had a long history of sending unmanned probes into deep space. Back in the 1970s, the number crunchers noticed that a rare planetary configuration of Jupiter, Saturn, Uranus and Neptune would not be repeated for another 176 years. This set the stage for the spectacular Voyager flights of the 1970s that revealed a huge amount about the solar system. The lab's Pathfinder lander touched down on Mars in 1997 and the Cassini probe went into orbit around Saturn in 2004, transmitting extraordinary photos of its rings, and they were also involved when the Huygens lander reached the surfaces of the planet's largest moon, Titan, in July 2004.

In the past forty years, the lab has been strongly affiliated with NASA, dispatching unmanned probes to seven of the planets and many of their moons. The lab is a nonprofit, government-funded research institute managed by the California Institute of Technology. One of the most famous images of the birth of a star was taken by JPL's planetary camera on NASA's Hubble space telescope in 1995, showing the startling view of the Eagle Nebula.

But the boffins have built up a fount of expertise vital if human flight is to continue beyond the Moon. It is their expertise that found out that the Martian winds could throw a spacecraft out of its orbit and that additional thrusters would be needed to counteract this.

Recently the lab has been planning GRACE – Gravity Recovery and Climate Experiment – a twin satellite designed to measure the Earth's gravitational field and the impact on the oceans. So Tito was only one celebrated individual from a remarkable organisation.

Tito took his Mariner software from JPL, where he worked for five years, and used it in investment management for pension funds, founding Wilshire in 1972. The company was at the forefront of using computers for investment and pension analysis. In some cases it was at least ten years ahead of the actuaries and accountants who adopted similar systems. The Dow Jones Wilshire 5000 Index, the broadest measurement of US securities, has become a recognised tool for thousands of businesses, making Wilshire billions of dollars in the process.

Faranetta continued: 'He took this software, used for calculating the probability of orbits, and he used it as a risk assessment tool for making investments. He has done very well. But it was his dream to fly in space and he fulfilled that dream with Space Adventures.'

He recalled that Dennis Tito was a stubborn man. 'When he first approached us he wanted to fly on the Mir space station. But it was clear to us that it was heading for the Pacific Ocean. [The 135-tonne satellite, which had been in space for fifteen years, was brought down from space in March 2001 in a programmed crash into the Pacific.] So, after some persuasion, he decided to go ahead with flying to the International Space Station.'

Dubbed the Cosmic Capitalist, Tito paid a reputed $18 m and undertook extensive training for six months at Star City. The ISS might be an international partnership involving the Russians and the Americans, but there are distinct modules and cultures in place. NASA was not going to approve of Tito's ride into space. But one of the legal agreements was that the Russians could select and train their candidates for flight. It almost sparked a strike with Soyuz cosmonauts anxious to fly, and there was extended hostility with NASA before a last-minute agreement that Tito would agree to pay for anything that might break – and not venture into the US sector. But cosmonaut Tito blasted off on 28 April 2001 from the Baikonur cosmodrome in Kazakhstan, generating massive publicity for space tourism.

Tito's mission in a Soyuz craft created a welter of interest in the new International Space Station. After seven days, twenty-two hours and seven minutes in orbit he returned to Earth safely and committed himself to sending more paying tourists into space.

The self-righteous *New York Times* criticised Tito for inventing 'the most offensively elitist form of eco-tourism yet devised' – although *Time* magazine retorted: 'Sorry, Dennis Tito has invented the most democratic from of ultra-capitalism yet devised. Let people pay what the market will bear to live out their fantasies.'

Tito remains unrepentant: indeed he has become a flag-bearer for space tourism and cheaper suborbital flights. 'If we're really going to see significant advances in the next hundred years in human space flights, I think it's going to happen from the private sector ... if there's enough demand there will be money from the private sector to make it happen,' he said at the Space Frontier Foundations' symposium, 'In Search of 2001', in Los Angeles in October 2001.

Tito admitted that one of his fears was not the blast-off from the launch pad but space sickness – or space adaptation syndrome. 'I had a small bout of space sickness the first day. It was a non-event. It came and went. It wasn't like being on a boat and being seasick all day. The rest of the mission was pure euphoria. The time I spent on the station was the most enjoyable period of my life. I slept like a baby and floated for eight days,' Tito said to Leonard David, the space writer of Space.com in October 2001.

Tito was followed into space by Mark Shuttleworth, a dashing young South African entrepreneur who now lives in London. Born in September 1973, he was a dot.com millionaire who sold his e-commerce security web-server company, Thawte, at the height of the stock market boom in December 1999 to VeriSign. The price was a handsome $575 m. Mark was smart enough to see that the burgeoning Internet needed some kind of specialised digital security that could authenticate certificates and transactions. So four years earlier, this 22-year-old computer-gaming fanatic set up Thawte in his final year of study at the University of Cape Town, where he was finishing a degree in finance and business information systems. His legacy lingers on as VeriSign remains at the heart of transactional security on the worldwide web today.

'By 1999, I was in a position where I could do anything, so I asked myself: What is the one thing you want to do before you die? The answer that came back immediately was to go into space, taking a step down an inevitable path that we as a species have to follow,' he said in an interview with *Air & Space* magazine in November 2006.

His experience – as a young Generation X game-player – was very different to Dennis Tito's. On 25 April 2002, he flew as a member of the Soyuz TM-34 mission, launched from

Baikonur in Kazakhstan, which docked with the ISS. He spent eight days working in space conducting science experiments for South Africa on AIDS and the human genome, and enjoying the weightlessness. In total, he spent 9 days, 21 hours and 25 minutes in space.

'Mark is one of the first people to fly into space before he was thirty. He actually scored the highest in his cosmonaut testing. He impressed the Russians with his ability to pilot the Soyuz spacecraft in the simulator. He had very high scores in the simulator, including docking it. He is a very capable young guy,' said Faranetta.

The third space tourist was a material scientist from Princeton, New Jersey. Greg Olsen struck it rich in high technology in the 1990s when he sold Sensors Unlimited, which makes fibre-optic devices. His moment of inspiration came while sitting having his regular coffee in a Starbucks, reading about Mark's mission in the *New York Times*.

'Basically we spent the time scouring the Earth for individuals. But here is a man who lived two miles away from where I stay, and we know the same people. He's an unassuming and regular guy,' said Faranetta.

Born in Brooklyn in New York, Olsen's father encouraged him to go to college instead of joining the army. Here he excelled and went to work for RCA, the electronics and recording company, pioneering the technique of growing ultra-pure crystals in a vacuum. He left and set up his business, selling it at the top of the dot.com market for $700 m – after the crash, he bought it back for only $7 m. Then he resold it again to Goodrich while he was waiting to go into orbit with his Soyuz capsule colleagues, cosmonaut Valery Torkarev and Bill McArthur of NASA.

Olsen made his first trip to Russia in October 2003 to prepare for a space flight two years later. But his $20 m mission was almost cancelled when Russian doctors at Moscow's Institute of Medical and Biological Problems found a spot on his chest X-rays. It took nine months of extensive tests and affidavits from US physicians to persuade the Moscow doctors – used to conducting rigorous tests on lean and super-fit cosmonauts – that an out-of-condition man in his fifties could fly.

Olsen agreed to release all of his medical data. He had been a smoker in the early 1980s, which damaged his lungs. 'So we had to do a very rigorous qualifications programme for Greg – and we learned a lot in the process. We now have very detailed medical information which will help future space tourists. It has been very helpful. It is the first medical paper written where you actually know who the individual is,' said Faranetta.

The intense cosmonaut training at Star City helped Olsen as he was returning to Earth. The commander of the Soyuz capsule bringing him back to terra firma shouted, '*Kislorod*' (oxygen) at him as the three crewmen sat jammed in the re-entry vehicle. The craft had developed a leak, causing a loss of pressure, and Olsen was the only one who could reach the oxygen valves that would save the lives of Olsen, commander Sergei Krikalev and NASA's John Phillips making the return from the ISS. His training kicked in immediately and Olsen was able to ensure everyone was given vital oxygen. The leak turned out to be minor but he said later, 'I could just tell by watching Krikalev that he had everything under control, and his control gave me confidence.'

Now the next space tourist scheduled to go up is Charles Simonyi, aged 59, who worked with Microsoft and developed Microsoft Word and Excel. The Hungarian-born American, listed as the 746th richest person in the world by *Forbes* magazine, faces the same rigorous regime before he can enjoy his trip into space in 2007.

A few weeks after her momentous space flight, on 20 October 2006, Anousheh Ansari flew in with New Mexico's Governor Bill Richardson on a private jet as a special guest at the Wirefly X Prize Cup at the tiny Las Cruces International airport in southern New Mexico. She arrived with husband Hamid and sister Atousa Raissyan to a hero's welcome, feted by the thousands gathered for the two-day event that attracted 20,000 visitors – including hundreds of noisy school kids fired up by big bangs, rocket bikes and a daredevil wearing a rocket belt. Her message should resonate with every politician and government policy wonk in every advanced industrial nation: encourage all young people to become excited about space; encourage them to take an interest in science and work hard with maths. The more savvy are listening.

'I am keen that we find ways to travel beyond our planet and explore the rest of the universe. I cannot think of any other organisation better than the X Prize Foundation that is going to make that a reality,' she said in an interview with the author.

'It is more than one person's job to keep this excitement going. I would like to see NASA, the European Space Agency, all the space agencies involved. They have immense resources that are not being used efficiently, so I am hoping to work together with them and, using my access to the kids, I want to promote space and make it a long-term inspirational plan for all young kids to be interested in space and learn more about it. I want more kids to go into the fields of maths and science so we can have a new generation of great scientists who can build better spaceships and go farther and solve some of the Earth's problems.'

But Anousheh is wealthy and can do much more. The Ansari family, through its Prodea investment company, is now actively supporting the Space Adventures suborbital programme, with the carrying plane in development. The Explorer, which will be launched from the United Arab Emirates, has nearly 200 reservations, each willing to pay $102,000 for the ride of their lives.

'We are getting people who have bought tickets with Virgin Galactic who have also bought tickets with us. Essentially, they are hedging their bets; depending on who makes it, they will have the first seat,' said Faranetta.

So now there is growing competition from commercial companies – but the response from the monolithic agency that has dominated and directed America's vision of space for over 45 years has been very interesting. Very interesting indeed.

6. NASA WAKES UP AT LAST

This is the tale of the arrival of space tourism, not the technical wizardry of bygone rocket projects. But to understand why the world is changing, we need to look at how commercial space is only now opening up for business.

Firstly, NASA is contrite. It admits it got it hopelessly wrong about paying passengers – and it wants to make amends. If space tourists wish to go and experience our fragile Earth or sample weightlessness – even for a short while – then it now wants to be part of this.

This is a dramatic turnaround from a government-funded organisation renowned for its arrogance and intellectual superiority. Even the Russians have shown the kind of entrepreneurial zeal more akin to American capitalism. As we have learned, they have been willing to take a handful of paying astronauts to the International Space Station. Slowly but surely, the space mavericks working on the fringes, the dot.com millionaires who dabbled in amateur rocketry and the computer-game designers who have netted fortunes are being brought into the fold, and given a seat at the table. There is even talk now of a 'global open architecture' – like a Linux operating system for space. This is unprecedented.

Chris Shank is NASA's Director of Strategic Investments and he has seen the light about space tourism. He is the budget guy who was previously a special assistant to Dr Mike Griffin, the Administrator of NASA, who took up his position in 2005. Today, Shank is trying his best to win new friends. NASA's budget for 2007 is $16.8 bn, which amounts to 0.6 per cent of the whole American federal budget. It is a colossal amount, worth more than the GDP of over half the nations of the world. And NASA calculates its costs for its plan to return to the Moon to be $24 billion from 2007 through until 2011.

Shank sat in on a panel discussion in New York with a group of senior players in the commercial space conference in October 2006. He was polite and expressed a sense of optimism. But what has changed is that NASA now accepts that commercial space flight has a role to play in the future of space exploration. And he talked animatedly about new areas of co-operation.

'For us to complete the International Space Station, to go back to the Moon . . . and then go on to Mars is going to require commercial and international investment. Not simply NASA. We cannot do it ourselves. We have come to the conclusion that the Apollo era of funding was not sustainable,' he told the gathering.

He explained that NASA's work had been unfairly characterised as 'Flags and Footprints'. 'We are in a new era now,' he declared. 'That's what our vision of space exploration is about – to open it up to more folks.'

So who are those folks?

NASA is working on heading back to the Moon and then on to Mars but 'we can leverage the work the new entrants are doing over the next years', continued Shank. The new guys are serious. And seriously rich too. One or two have mega-fortunes that can compare with NASA's income.

In January 2004, President George W Bush gave the go-ahead for NASA to take human beings back to the Moon and set up a permanent lunar base as a stepping stone to Mars. It was a presidential option – not a mandate.

This is the Constellation Programme. Dubbed Moon 2:0, NASA is now designing, testing and evaluating the systems for this new generation of exploration. The chief component is the Ares I rocket, which will carry human beings safely and reliably to the Moon and then to Mars. The Ares I will take the crew, while all the heavy lifting will be done by the Ares V rocket, and a new Orion capsule is planned as the home to astronauts. The Crew Exploration Vehicle must be ready by 2014, with an extended life until 2020. The intention is to carry four people for sixteen days. It's brave and exciting stuff.

In terms of rocketry, it is hardly ground-breaking. The Ares V is a two-stage vertically stacked launch system. The reusable boosters will use space-shuttle technology. The core propulsion is liquid oxygen and liquid hydrogen – really an upgraded version of the current Delta IV, developed in the 1990s by the US Air Force. It's not major innovation but it will do the job. But NASA will not get humans back on the Moon before 2018.

And, Houston, we have a problem. Money. That old chestnut was again rearing its head. So a $104 bn plan to refocus NASA on a human space flight beyond low-Earth orbit was being downplayed. Shortfalls in funding for the space shuttle and the International Space Station and the Hubble space telescope were testing NASA's ability to go for the Moon. In Congress, supporters were quick to point out the flaws in the financial argument. The budget, quite literally, was bust. Mike Griffin, the former head of the space department at the Applied Physics Lab of John Hopkins University and a former chief engineer, faced shedding 15 per cent of his staff by the middle of 2006.

The budget figures were enormous. How would the American public approve?

Then President Bush became embroiled in a more immediate numbers game. In the summer of 2005, American military forces were still deeply committed in Iraq and Afghanistan.

Every day the death toll was rising as suicide bombers targeted Americans. The body bags were being flown back to the States, and there were demands for more military hardware and troops. The budget was haemorrhaging.

Sherwood Boehlert, the chairman of the House of Representatives science committee, and a Republican to boot, declared there was 'simply no credible way' to finance the expedition to the Moon unless NASA asks Congress for more money.

The Bush administration dug in its heels and refused.

'Whether such an increase is a good idea in the context of overall federal spending at this time is something neither Congress nor the administration has yet determined,' said Boehlert in an interview in *Flight International*.

It had the distinct appearance of a retread of the Apollo mission and was fantastic fodder for those conspiracy theorists and Internet bloggers who still believed the Moon landings were an elaborate hoax. Ask many smart young people in their late twenties and early thirties and they find it very hard to believe that NASA could ever have taken astronauts to the Moon in the 1970s, yet done nothing as spectacular since then. To them, this all stinks of rotten fish. Even Griffin had to admit the limitations of a project aimed at returning to the Moon in 2018. 'Much of it looks the same as Apollo but that's because the physics of atmospheric re-entry haven't changed recently,' retorted the NASA chief.

So NASA had overslept – but its wake-up call has been loud and vibrating. While NASA had been co-operating with the Russians, Japanese and the European Space Agency, it badly needed some entrepreneurial flair. Now it has some.

In the summer of 2006, it awarded contracts to two space companies for commercial orbital transportation services. One to SpaceX, the company founded by Elon Musk, the billionaire founder of PayPal, and the other to the Kistler Aerospace Corporation for their K-1 RocketPlane.

A few weeks before Shank's comments, on 18 August 2006, NASA selected Musk's Space Exploration Technologies (SpaceX) to send a commercial rocket up to the International Space Station in late 2008, around the same time as

SpaceShipTwo is completing its own test programme. If the California-based company is able to demonstrate its capability of supplying cargo to the ISS after the retirement of the space shuttle, this will be a major change of direction for commercial space operations.

The agreement means SpaceX will undertake three flights using its Falcon 9 rocket carrying a Dragon reusable capsule that is a mix of Apollo and Soyuz. Meanwhile the Kistler mission is to develop a low-cost reusable aerospace vehicle, designed to deliver payloads to orbit and also take cargo and supplies to the ISS.

Both are a kind of FedEx of space.

'We're taking the gamble on commercial space,' declared Shank. Increasingly, it looks like a good bet.

Fresh-faced Elon Musk was born and grew up in South Africa but left home at seventeen to avoid compulsory service in the South African Defence Force. Musk's early career was in a number of advanced-technology industries, from high-energy-density ultra-capacitors at Pinnacle Research to software development at Rocket Science and Microsoft. He has a physics degree from the University of Pennsylvania, a business degree from Wharton and came out to California to pursue graduate studies in high-energy-density capacitor physics and materials science at Stanford. He is one incredibly smart cookie.

He became the archetypal dot.com millionaire who struck it rich. In 1999, at the height of the boom, he sold his Internet company, Zip2, a creator of online city guides, to Compaq for $307 m in cash. It had only been in existence for four years. He then created X.com, which became the hottest property in Silicon Valley after morphing into PayPal and going public on the NASDAQ exchange in 2002. The online auction company eBay has also been one of the sensations of the Internet era, a site where millions of consumers around the world can buy and sell almost anything that's legal – and some things that aren't! The beauty was a new method of electronic payment for people to transact as safely and effectively as possible. Musk's PayPal was the business, and its success was sure and rapid. And for Musk it was a gold mine when he sold it to eBay for $1.5 bn in October 2002.

Musk sees a clear direction for his business: 'Once our Falcon 1 has a few flights under its belt and the satellite producers have time to adjust, I think it is quite possible that there will be more flights per year of Falcon 1 than any other vehicle in the world. It is worth noting that the Falcon 1 is the only semi-reusable rocket in the world, apart from the space shuttle.

'However, reusability is not currently factored into the price. As we refine that process, the cost of Falcon 1 will decline over time. As far as reliability is concerned, the Futron corporation, which is used extensively by NASA and the FAA, concluded that Falcon 1 had the second highest design reliability of any American rocket,' he told a House of Representatives Space and Aeronautics Committee. Indeed, it was tied with the most reliable version of the Boeing Delta IV and Lockheed Atlas V. The highest design-reliability rank was held by the Falcon V design, which Musk says is the only American rocket that can lose any engine or motor and still complete its mission.

The Falcon V is a medium-lift rocket designed to carry people as well as much larger satellites. As such, says Musk, the design margins will meet or exceed NASA requirements for manned spacecraft. 'My hope is that this vehicle will provide the United States with an all-American means of transporting astronauts to orbit and ensure that we are beholden to no one once the shuttle retires.'

But Musk wasn't the only star with cash.

Rocketplane Kistler's K-1 programme has also been chosen by NASA's Orbital Transportation Services. Rocketplane and Kistler Aerospace Corporation joined forces in February 2006 as a new space transportation team. Kistler Aerospace had only just emerged from chapter 11 bankruptcy protection in March 2005, and had debts to pay. Here were two of the most entrepreneurial and technically advanced companies in the new space business community putting their resources together. It showed how tight the market was for investment.

Space projects have the ability to burn cash like rocket fuel. The Rocketplane Kistler team is aiming for both suborbital and orbital commercial space transportation services for passengers and cargo through a fleet of highly reliable reusable aerospace vehicles.

One of its suborbital vehicles, a modified Lear jet, is hardly revolutionary. Frank Bauer, overheard speaking in New Mexico, said there wasn't any point in building something from scratch and that they would be adapting existing technology. 'There's no point in us reinventing the wheel. What we have is a derivative of the RS-88 from Rocketdyne, but one of the spin-offs when Rocketdyne was bought by the world's leading engine companies, Pratt and Whitney, was that the guys who worked on that programme set up their own engine company in California called Polaris Propulsion,' he explained to the author.

David Crisalli, head of Polaris Propulsion, started his company in Oxnard, California, in 1997, before joining Rocketdyne. Now Crisalli and his Polaris associates are taking the RS-88, capable of up to 50,000 lb of thrust and built for use on Lockheed Martin's Pad Abort Demonstration vehicle, and building a nozzle for re-entry. In 2003, NASA tested the RS-88 in a series of fourteen hot-fire tests. 'We are looking at a derivative of this existing engine,' added Crisalli, 'we are not dealing with the complete engine risk programme. It's powered by Lox (liquid oxygen) and kerosene with 36,000 lb thrust.'

The Rocketplane Kistler company, working with Andrews Space, has already been test-firing its engines. The company has one RS-88 on agreement from NASA's Johnson Space Center, testing with the existing engine and then with a re-entry nozzle.

There is still another big player, though, as an unlikely impresario has shaken his gold dust in the air. Robert Bigelow, the Las Vegas property magnate, launched a test vehicle 340 miles (547 km) into orbit on 12 July 2006, when a Genesis I rocket was launched from Yasny, the ISC Kosmotras Space and Missile Complex in the Orenburg region in Russia. Its payload was a massive grey and blue inflatable structure, which was inflated like a massive sausage. Bigelow, who loves a good PR angle, said, 'It feels just like becoming a father. It's our little baby that's up there.'

The Genesis I spacecraft is only a one-third size prototype and its initial crew were cockroaches and Mexican jumping beans to test whether life-support systems were working. But the grand idea is to build a hotel in space.

Some are highly sceptical. John Loizou, an aerospace engineer with the Vega Group, put it into some kind of context when he told the *Guardian* newspaper in Britain: 'Certainly he's got grand ideas and I really want it to happen. Building prototype hotels at the moment just strikes me a little bit like running before you can walk.'

And with so much debris floating around in Earth's orbit, there was a real danger that something sharp could puncture the inflatable Bigelow dream.

This burgeoning market was also inspiring the driven Russians, now catching up quickly when it comes to embracing Western free-market economics. Nicolai Sevastiyanov, the president and general designer of RSC Energia, said his company wanted a slice of the space-tourism pie. Speaking through an interpreter, he told the International Astronautical Congress in Valencia in October 2006 that there were parallels between government and private investment and the development of Siberia's aviation industry in the 1930s. 'Some people were saying, "Why do we need to do this? It makes absolutely no sense because nobody can live in the polar regions." But later on, very large deposits of gas, oil and other mineral resources were discovered in Siberia and aviation became the major technology for exploration and development of these resources.'

So the Russians have been pillaging the rusting Soviet scrapyards to find rockets that have been stockpiled since the 1960s. They found 70 first-stage N-1 Moon launch vehicles – all in workable condition – and available for use on a new, reusable vehicle called the Kliper. This would be space tourism, Russian-style, run like an airline.

'If you are a pilot, and can fly an aircraft, you shall be able to fly the Kliper. Instead of tens, the army of space tourists will rise to hundreds,' Sevastiyanov said at MAKS 2005 Moscow air show. 'We will make the International Space Station a spaceport. This may lead to a business similar to the airlines. Like Boeing and Airbus, we will be able to sell space vehicles to other countries.'

It's a bold plan – which has attracted interest from the European Space Agency – but there has been some scoffing at what is dubbed Spaceflot, a disparaging reference to the Russian national airline Aeroflot, not previously known for its levels of comfort and service.

What was clear was that these new companies – however imaginative – were still deploying ancient space technologies. There was nothing innovative about old missiles from the Soviet era, and ferrying half a dozen exceptionally rich people once or twice a year was not going to open up a larger market. There was a need for some drastic change – indeed, a paradigm shift. And this would require a new way of taking paying passengers into space.

Showbiz was one thing; but concrete ideas were another. Griffin, the head of NASA, began looking at reasons and strong arguments for returning to the Moon. Increasingly, he was up against a 'Why go to the Moon?' lobby challenging this decision. And there was a growing environmental campaign against using up Earth's resources by going into space. Griffin wanted real answers. He said in an email to key NASA insiders: 'We've got the architecture in place and generally accepted. That's the "interstate highway" analogy I've already made. So now, we need to start talking about those exit ramps I've referred to. What ARE we going to do on the Moon? To what end? And with whom? . . . Now is the time to start working with our own science community and with the internationals to define the program of lunar activity that makes the most sense to the most people.'

So new ideas and fresh thinking were desperately needed. And Molly Macauley, of Resource for the Future, speaking on 16 July 2004, raised some pertinent points in a response to the

US House of Representatives science committee. Resources for the Future is a nonpartisan research organisation in the United States which conducts independent analyses of issues concerned with natural resources and the environment.

Macauley's area of research is space policy with a focus on economics, including space transportation. 'We have searched for the silver bullet that would propel our nation back into space by way of the shuttle and space station for the multiple pursuits of scientific exploration on the one hand and a vibrant commercial space industry on the other. There is no lack of ingenuity in ideas for both of these goals. But critics of NASA's plans – regardless of the specific details involved – assert that they take too much time and money away from more pressing societal needs.'

It remains a perpetual complaint. With millions around the world dying each year through civil war, famine, hunger and AIDS, tens of millions of children perishing through cholera, tuberculosis and malnutrition, and with a global fight against terrorism, the vast spending on space travel remains an emotive issue. Even in the United States' homeland there have been vociferous calls to cut back – especially after the destruction of New Orleans in the aftermath of Hurricane Katrina in 2005.

'Prizes, although not a silver bullet for invigorating enthusiasm for space or elevating its priority in spending decisions, could nonetheless complement government's existing approaches to inducing innovation – procurement contracts and peer-reviewed grants,' she said.

'Even if an offered prize is never awarded because competitors fail all attempts to win, the outcome can shed light on the state of technology maturation. In particular, an unawarded prize can signal that even the best technological efforts aren't quite ripe at the proffered level of monetary reward. Such a result is important information for government when pursuing new technology subject to a limited budget.'

Research grants rather than prizes typically financed studies of rockets – although even research grants were rare in the early decades. Konstantin Tsiolkovsky, Robert Goddard and

Hermann Oberth – the fathers of space travel – worked independently in self-financed home-based or academic laboratories. Tsiolkovsky received a grant of 899 roubles in 1899 from the Russian Academy of Science. Goddard, after making multiple requests (with the support of Charles Lindbergh), was given grants of $5,000 and $3,500 from the Smithsonian Institution during 1917–20. In 1927, some forty years after the first articles on rocketry had been published, Robert Esnault-Pelterie, an airplane inventor, and Andre Louis-Hirsch, a banker, set up a 5,000-franc prize. The prize was to be awarded annually to the author of the most outstanding work on astronautics. The heyday of prizes was from about 1900 to 1917 – two decades in which aviation feats filled the news, meeting a seemingly insatiable appetite for daredevils and barnstormers.

Eighty years later, the X Prize proved to be the impetus. Now NASA is also prepared to put some of its government money up to reward innovation. For example, there is now the $2 m Space Elevator challenge for lab scientists and university students, where the idea is that in the future it might be possible to push an 'up' button and payload will be pinged into space. A space elevator is a system based on a super-strong ribbon – able to bear its own weight – going from the Earth's surface to a point beyond geosynchronous orbit. Like a massive elastic band in space, it is held in orbit by a counterweight on a space station. As the Earth rotates the tether is held taut and so vehicles would climb up the ribbon, powered by a beam of energy projected from the Earth. It is an utterly mind-blowing idea but one that is seriously focusing many minds – especially those using new carbon nano-tubes. NASA says it needs a material that is 25 times stronger than anything currently in existence – this is a greater leap than from wood to metal, a very tall order indeed.

And young entrepreneurs such as John Carmack have come along to take up the Lunar Lander Challenge, another NASA Centennial project. The challenge is intriguing, set up to encourage external innovators prepared to build a vehicle capable of ferrying cargo or humans back and forward between lunar orbit and the surface of the moon. The $2 m purse is enticing – but NASA wasn't sure if anyone would really be able to build a suitable rocket-propelled vehicle before 2010. At the rocket show at Las Cruces International airport in October 2006, Carmack, guiding his vehicle using a computer-game joystick, managed to raise his Armadillo Aerospace rocket-propelled lunar-landing vehicle into the air. Like a weird UFO, it drifted sideways for a few hundred metres before returning back to Earth. To win, he needed to repeat the feat within two hours, but it wasn't to happen this time. But the fact that this computer-games nerd had cracked a major technological challenge in six months proved things were really happening. And faster than anyone could have predicted.

7. THE X PRIZE GUY

As NASA was learning, the world loves prizes. And success in the world of aviation has been built on winning trophies. It was two Frenchmen who both loved the thrill of flying – Jacques Schneider and Raymond Orteig – who were the posthumous inspiration for a commercial space prize.

Jacques Schneider was a French industrialist and an early adopter of the plane. He was a fanatical balloonist, a veritable Richard Branson of his day, who set the balloon altitude record of 33,074 ft before a serious accident cut short his career. He continued his passion by supporting various competitions and aero clubs. This was the time of flimsy flying machines held together with rope, canvas and wood, where there was exhilaration and ever-present danger.

As a race referee at a Monaco meeting in 1912, he noticed that seaplanes weren't able to fly as fast as planes taking off from runways. Yet, living on the Mediterranean, he could see that seaplanes were a solution for long-range passenger services and thought a contest might spark some fresh thinking. On 5 December 1912, at the Aéro-Club de France, he offered a trophy for a seaplane race and proposed a course of at least 150 nautical miles. This was the Schneider Cup, or the Coupe d'Aviation Maritime Jacques Schneider.

The magnificent trophy cost 25,000 francs to make and to secure it a pilot needed to win three races in five years – and also pick up 75,000 francs in prize money. Like the America's Cup sailing race, each race was to be hosted by the previous winning country. The races were to be jointly supervised by the Fédération Aéronautique Internationale and the Aero Club in the host country.

Each club would be permitted to enter up to three competitors with an equal number of alternates. Crowds of over 250,000 gathered to watch the Schneider Cup races, proving a keen public interest in this type of competition. It was a tough battle – and for years no one was able to secure the prize outright.

Meantime, in the grand salons of one of New York's most famous hotels, the Brevoort, another aviation prize was being hatched. Brevoort House had been the sumptuous stopping place for titled Europeans who sailed across the Atlantic. In 1902, it was renovated by its new owner, Raymond Orteig, a 32-year-old maitre d' from Bearn in France, famed for his knowledge and love of the best French wines. Each year he embarked on a wine trip but yearned for a quicker way to cross the Atlantic back to Bordeaux. In 1919, after the carnage of the First World War, he offered the $25,000 Orteig Prize for the first nonstop transatlantic flight between New York and Paris. The prize was originally offered for a period of up to five years, but the deadline was extended. It was this prize that was captured by Charles Lindbergh when he landed his *Spirit of St Louis* in Paris on 21 May 1927 – thirty-three and a half hours after setting off from Roosevelt Field on Long Island, in the United States. Lindbergh's dramatic flight changed the way people viewed air travel.

By 1927, the Schneider Cup was still up for grabs – but it would produce one of the flying marvels of its age, the Supermarine Spitfire. In 1925 the UK's air ministry formed a racing team at Felixstowe, Suffolk, and commissioned the designer Reginald Mitchell to develop a successor to the Sea Lion II. The result was the Supermarine S5, which won the race in 1927, followed by the Supermarine S6, which triumphed for a third time in 1928. Supermarine collaborated with the engine manufacturer Rolls-Royce to develop a powerful 12-cylinder engine for the S6. That plane reached a winning speed of 328 mph. Mitchell and his team then modified the design of the S6 into the S6B for the 1931 race. It won the event for Britain and the right to keep the coveted trophy. A few weeks later the S6B broke the world speed record by flying at 407 mph, a record that remained unbroken for 14 years. The legendary Spitfire evolved from the original seaplane that won this prize.

Both tales fired up an American, the exuberant Peter Diamandis, who hoped that a new prize would do the same for commercial space travel.

'Peter is truly the Raymond Orteig of our time,' said his long-time friend and partner Gregg Maryniak. Diamandis is an intense, impeccably dressed son of Greek immigrants, a man so obsessed by space that even his mother jokingly wonders if her son carries an extraterrestrial gene. But he knew one thing about prizes – they were the catalyst for technological changes, even game-changing leaps forward.

Diamandis's idea was a straight lift from the early history of aviation when, between 1908 and 1915 – the heyday of privately sponsored competitions for distance, elevation, and speed – nearly forty individual prizes fostered great technological leaps. The first prizes were for flights of 25 m and 100 m, then for over 1,000 ft in elevation. Subsequent prizes were for longer distances, higher elevations and faster times. Prizes were offered by private individuals and companies. In addition to Orteig and Schneider, the list included: Ralph Pulitzer, the son of newspaper publisher Joseph Pulitzer; James Dole, a Hawaiian planter; Eduoard and Andre Michelin, the executives of the Michelin Tyre Company; and James Gordon Bennett, the publisher of the *New York Herald*. The French especially loved prizes, with the French Aero Club and the French Champagne industry supporting competitions, while the Harvard Aeronautical Society, the Daniel Guggenheim Fund, the Harmsworth *Daily Mail* in London and the *New York World* all pitched in. Exclusive stories about adventurers sold newspapers and magazines.

Armed with this knowledge, Diamandis began his quest. It took time, but people began to sit up and listen to him when he spoke persuasively about aviation history. Although he has a large ego and loves the spotlight, he is likable and people trusted him. Friends and colleague were prepared to back him, bail him out even, when others had ridiculed his notion of offering a $10 m prize for the first privately financed passenger craft to soar 62 miles through the atmosphere and return safely to earth, then repeat the feat. This was the X Prize.

'I'm proud that the X Prize put suborbital flight back on the map. Before our announcement in 1996, no one was thinking about this as a market or a useful technology. The X Prize basically defined space as a hundred kilometres up in the air and three people as a new class of launch capability,' he says.

Over the years, Diamandis has been an evangelist for commercial space, organising space conferences, setting up websites and starting foundations to promote space travel. He founded the International Space University, which started as a summer school and now has a permanent campus and staff in Strasbourg, France.

Having gained a medical degree from Harvard University in Boston Diamandis went on to take an aerospace engineering degree at Massachusetts Institute of Technology. Instead of joining NASA he decided to start his own rocket company and co-founded the Zero Gravity Corp, which eventually gained federal aviation approval to take the paying public on weightless flights aboard a specially modified Boeing 727-200.

'Probably my greatest attribute is my stubbornness ... my unwillingness to give up! Even in my other companies you can see that. My Zero-G Experience company where we give people weightless flights took eleven years to get FAA approval!' he says.

Diamandis needed that stubborn trait in spades. 'I probably

pitched the X Prize idea to two hundred chief executive officers over six years from 1994 and with every presentation I convinced myself over and over that this was a viable concept and one that would succeed. I kept telling myself it was just a matter of finding the right person.'

He says he never knew finding such a visionary person would be so difficult.

'I thought that Sir Richard Branson was perfect for it, but even he wasn't ready to take the risk in the early days. I got close with Fred Smith of FedEx and with a few of the large automotive brands. It really needed an individual entrepreneur willing to take the risk. The Ansari family was perfect.'

So how did it happen?

'I read about the Ansari Family in *Fortune* magazine in 2001 and visited them in early 2002. In Anousheh's write-up in that magazine she specifically said that she wanted to experience a *suborbital* flight into space . . . I must have read that line over three different times. It was their first meeting after they returned from vacationing in Hawaii. They truly are wonderful people and besides having them come aboard as our title sponsors, they've become life-long friends.'

Anousheh Ansari recalls her meeting with Peter. 'Two of our family members have a passion for space travel, myself and my brother-in-law Amir, and when we found out about X Prize and how suborbital might be the way for the future, we were very keen to support it,' she says.

Peter went to see Anousheh and Amir just after they had sold their company for $750 m, so the family weren't short of cash. 'We were just back after a six months break from selling our company. My assistant told me that there was a guy who was very persistent and wanted to talk about something to do with space. She knew how much it interested me,' Anousheh tells me in an interview in New Mexico.

Diamandis went to Dallas and gave a big PowerPoint presentation to Amir and Anousheh. 'He had us sold on the first two slides but we let him go through it and tortured him a little bit. But we looked at each other across the table and we knew this was definitely the right way to do it.'

Not only did Peter have his $10 m prize fund, he now had the enthusiastic Ansaris firmly bought in – and this relationship would soon develop and prosper.

So does Ms Ansari feel that the X Prize and Peter Diamandis's involvement can have a more positive impact on the environmental questions being raised as an increasing number of people take commercial space flights? 'I think there is a lot more pollution from other sources on Earth. But definitely one of the areas the X Prize Foundation is focusing upon are breakthroughs in new propulsion systems. We are trying to encourage new sources of energy. And we hope that any type of vehicle or spaceship that uses fossil fuel can start using environmentally safe fuels. We are trying to be a positive force behind innovation.'

Diamandis, now in his mid-forties, remains deadly serious about his dreams. And these dreams go far beyond mere commercial space travellers arriving in suborbit. He has visions of humans living in space, exploring the stars, and of eventually colonising them. 'I believe people now get the idea that we need to start with a near-term step that can generate a profit and then build from there instead of jumping straight to orbit,' he says.

It's why he is keen on the involvement of Virgin Galactic. 'Suborbital flight technology, on the other hand, is achievable now, at a reasonable cost, and will ignite the world's interest and enthusiasm for space.'

But Diamandis also makes a plea for more innovation and the appreciation of calculated risk. 'I think we are killing ourselves in the United States and on Earth as a species by becoming so risk averse. You cannot have breakthroughs without taking risk. The day before something is truly a "breakthrough" it is a "crazy idea" . . . like computing on silicon rather than vacuum tubes. If it's not a crazy idea then it's probably not a breakthrough but an incremental improvement. So how do we embrace and allow for crazy ideas? How do we allow risk? Governments fear Congressional investigations and large corporations fear having their stock prices plummet . . . so most breakthroughs come from individuals or small corporations. We must create environments where we are allowed to fail, allowed to risk.'

Recent, independent studies also support the financial viability of suborbital companies and the large market for suborbital flight. The Futron study in October 2002 projected that it would be a $1 bn market in ten years' time. But it was the success of the X Prize in the summer of 2004 that gave Peter Diamandis the impetus to keep going on. He deserves much of the credit for creating the competition.

8. WHERE THERE'S A WILL

The genesis of Virgin's space project came about during a lull in a Virgin around-the world ballooning attempt at Christmas in 1995. An illustrious gaggle of explorers was stuck in Marrakech waiting for the weather to change.

Because of the high-level climatic conditions, the expected jet streams required to take the Global Challenger balloon upwards were not appearing over Morocco and the flight had to be delayed. Sir Richard Branson had become fascinated by the jet streams and the currents in the thinner, upper atmosphere, and how they could be used for a hot-air balloon.

'We were sitting around one night in the Marriott hotel with Per Lindstrand, Buzz Aldrin, Alex Ritchie [the brilliant balloon designer], Richard and myself,' recalls Will Whitehorn, now president of Virgin Galactic.

The adventurers did what most adventurers do when they're off duty: swap tales of derring-do. And that evening there were some amazing tales exchanged.

It was extremely difficult to trump Buzz Aldrin, the man on the first Apollo mission to the Moon. Of all the astronauts who went to the Moon, Aldrin has remained most firmly in the public's consciousness – and popularity.

He was born in Montclair, New Jersey on 20 January 1930. His father, Edwin Eugene Aldrin, was an aviation pioneer and a student of rocket developer Robert Goddard. Buzz was educated at West Point, graduating third in his class in 1951. After receiving his wings, he flew Sabre Jets in 66 combat missions in the Korean War, shooting down two MiG-15s. In October 1963, Buzz was selected by NASA as one of the early astronauts, and in November 1966 he established a new record for extra-vehicular activity in space on the Gemini 12 orbital flight mission.

Buzz has logged 4,500 hours of flying time, 290 of which were in space, including eight hours of extra-vehicular activity. As back-up command module pilot for Apollo 8, the first flight around the moon, he significantly improved operational techniques for astronautical navigation star display. Then, on 20 July 1969, Buzz and Neil Armstrong made their historic Apollo 11 Moon walk, becoming the first two humans to set foot on another world.

Interviewed in a live broadcast at the X Prize Cup in New Mexico in October 2006, he was asked why, at 76, he was still fighting every day for the continued exploration of space.

'Well, I think it is so important to this country that we can maintain the leadership that we put together in such a short time. I think it is such an inspiration to young people today to be able to see that there is something that they can stand back and admire – that was done a number of years ago,' said the veteran astronaut.

He explained he was pulling together all the Apollo mission astronauts for a fortieth anniversary, beginning in January 2007 with a memorial for the Apollo 1 crew who perished on the launch pad. 'We had our challenges starting right from the beginning of Sputnik – and then Yuri Gagarin was the first to orbit the Earth, but even before that we saw dogs going into orbit and pictures taken of the back side of the Moon. And that's why all the craters on the back side of the Moon are all named after Russians.'

He was then asked why he had been pushing for many years – long before the X Prize Foundation – for regular folks to go into space, rather than simply a trained astronaut elite.

'Well, I had to deal with my own situations after what I did. After coming back from the Moon I decided to return to the air force. It seemed like a good idea. But because I didn't have a test pilot background, I chose a more professional and academic route – using informed analysis – by getting a doctorate degree from MIT. Then, after leaving the air force, I wasn't sure what I was going to do and the mainstay was what I knew most about and that was how to plan missions – and how to analyse trajectories and ways to go back to the Moon. I was intrigued by more novel and challenging ways of going into space. And I was trying to cultivate innovation and translate that into better ways of getting to Mars.'

Since retiring from NASA, the USAF and from his position as commander of the test pilot school at Edwards Air Force Base, Aldrin has remained at the forefront of efforts to ensure a continued leading role for America in manned space exploration. He created a master plan of evolving missions for sustained exploration called the Cycler, a spacecraft system that makes perpetual orbits between Earth and Mars. But he has also been available for other projects too, which is how he became a close friend of Sir Richard Branson.

Branson found Buzz's knowledge and insight fascinating. Later on that evening, Richard stroked his greying beard and turned to Buzz. 'Why do they launch rockets from the ground? Why don't they carry them up in the air in a balloon like we've been doing – and launch them from higher up?'

Buzz sipped his Diet Coke and said, 'Actually, we did.'

'What? When was that?'

Buzz took a moment to recall and launched into an explanation. 'Richard, what you've just described actually happened. The US government was experimenting in the 1950s with balloons flown off of our aircraft carriers in the

Pacific. They were launched high into the atmosphere with little rocket payloads, then fired upwards.'

'So what happened?' asked Branson.

Buzz laughed. 'Sputnik. It was a promising era but the Russians scared the living daylights out of the Americans. The experiments were shelved with all hands on deck for the Gemini project.'

Buzz explained the history of NASA's creation and the launching of ground-based rockets and he talked about the X-15 project driven by pilots. It was a bar-room tutorial from the best man to explain the history of American space flight.

Next morning the weather was much clearer. Tucking into omelettes and couscous for breakfast, Richard turned to Will Whitehorn and said, 'Oh, by the way, Will, you must make sure the Virgin brand is registered for space.'

Whitehorn went off and checked and found that by default when they had registered the Virgin trademark for aviation it also covered space. Richard also added, 'Keep an eye on space, because I think it is very exciting for the future. There must be a better way to do it than blasting off from the ground.'

Will Whitehorn was soon to become immersed in a new mission as the figurehead of Virgin's bid to take tourists into space. It was an appropriate task for one of the Virgin Group's most eloquent directors. Born in Edinburgh in 1960, Whitehorn lived in the well-heeled Grange area of the city. Like many British kids in the 1960s he grew up with a fascination for popular science and space.

'From the 1950s onwards, it was the dream that was presented to kids in magazines like the *Eagle* and children's picture books, such as the Ladybird series on the planets and space. There were fantastic cut-away pictures of massive Ferris wheels in space – which is still the logical way to do it. The rotating von Braun wheel housing a future space station was an iconic childhood image in comics like the *Eagle*.'

Whitehorn's parents used to take him to Edinburgh's Turnhouse airport, which had been an RAF base during the Second World War, but was now serving the growing capital.

By the 1960s, a glass observation tower and deck and a café had been added, which were a popular spot for lunches on a Saturday. 'We used to go regularly and I would spend my time with my nose pressed up against the glass watching the Vanguards, Viscounts and Tridents in the mid-1960s, which was the first jet aircraft I ever saw.'

His father Donald was an architect but had been a Royal Artillery officer during the war and he shared with Will the basic principles of ballistics and gunnery. It obviously worked: young Whitehorn became an excellent shot. 'From an early age I understood the concepts of trajectory and when I was young I learned to shoot at our weekend cottage in East Lothian.'

But space was also an interest. The mythology of space in the 1960s was that everyone was going to go there in the future. 'It was where our future as human beings lay. The Apollo missions seemed to confirm that,' recalls Whitehorn. Like so many Scottish compatriots, he became excited with ideas. At the age of twelve he was taken by his father to London to the Science Museum in South Kensington, where his grandfather's car, a hybrid petrol-electric engine used on London's buses between the wars, was an exhibit. Yet Whitehorn was held back from pursuing a career in science. 'I always wondered about that. I had a real problem with maths. It was only by forcing it through that I got there. But I was good at physics, chemistry and biology.'

He went off to study history and economics at Aberdeen University in the northeast of Scotland. 'I was reading the *Economist* at fourteen but my maths was a very tricky area. If it had been the age of the electronic calculator, who knows, it might have been very different.'

After graduation he took a temporary job working in the North Sea oil industry in Aberdeen. British Airways Helicopters were looking for search-and-rescue crewmen flying twin-engine Chinooks and single-engine Sikorsky S61Ns, the workhorses in the North Sea, and Will joined up. After six weeks training, in freezing water and rescue drills, he spent time flying as a crewman. But there were some serious accidents with a few people killed and Will decided it wasn't a safe career path.

He then became a graduate trainee with Thomas Cook, the global travel company. Ironically, one of his first projects was to find out if the tour company, set up by a Baptist preacher in 1841 as an eleven-mile railway excursion to take his temperance society from Leicester to Loughborough, had any liability from all the deposits they had taken for future space flights. But NASA's Challenger accident on 28 January 1986 – which killed seven crew members – put its plans on hold. So Whitehorn moved on to a bank working in financial public relations before joining Virgin in 1987.

He has worked for Sir Richard Branson for twenty years – a fierce loyalist who started as a press spokesman and became a confidant, friend and a wonderful foil for the high-profile entrepreneur. He has had a hand in almost all of Virgin's major projects over the last decade. It was Whitehorn's determination to help the UK's ailing railway industry that persuaded Branson to set up Virgin Trains and bid for railway franchises. It wasn't an easy proposition; the company had an abysmal reputation with passengers in its early days, perhaps unfairly because it had inherited ancient rolling stock, operating on clapped-out tracks on the West Coast main line. More recently, with billions of pounds of investment by the government in track upgrades, and Virgin's spending on tilting Pendolinos express trains, Virgin Trains has restored its reputation.

However, it would take a few years before Whitehorn's own space adventure would turn up. Peter Diamandis picks up the events.

'I had come up with the X Prize concept in 1994 and over the course of many years had presented it to numerous CEOs. I had heard that Richard Branson was interested in the subject so in 1997, about a year after our launch, I made it out to England to pitch to him. It would be the first of three attempts to get him to call it the Virgin X Prize. The first two times he didn't seem that interested in space flight and thought it was too dangerous and too immature. On the third attempt in 1998 it came close and was being proposed as the Virgin

Atlantic X Prize. We were to pitch it to the Virgin Atlantic board, but then the investment with Singapore Airlines came along and we took the back burner,' he recalls.

In late 1998, Peter Diamandis came to Holland Park to talk about the X Prize. Lori Leven set up the meeting but Branson was not available that day, so Whitehorn stepped in instead. He liked Peter and his infectious enthusiasm and picked up on the idea straight away.

'Absolutely great,' he thought. 'But . . .'

And there was a big but. 'We wanted to take the technology forward and build a business rather than sponsor a prize.'

Whitehorn told Diamandis he was supportive but that Virgin might participate in another way. They would keep an eye on the project and look at what was emerging – then make a move. Whitehorn offered a little bit of advice on insurance, explaining how Virgin had insured their balloon flights against success. 'If we failed then there would be a pay-out, something which Steve Fossett had done on his balloon flight as well.'

In March 1999, Whitehorn sent out a note to his London legal advisers, Harbottle & Lewis, asking them to register the company name: Virgin Galactic. Having set up the name, he began thinking about the next steps. 'Richard and I went out to the Mojave Desert to have a look at the Rotary Rocket project,' he recalls.

The Rotary Rocket Company was novel. Here was a single-stage rocket ship trying to deliver payloads into space at dramatically reduced costs compared to multiple-staged vehicles. The Roton Atmospheric Test Vehicle – the ATV – was built for the specific purpose of testing controlled landing and hover capability. Brian Binnie was the on-site programme manager and co-ordinator for the company.

The ATV used salvaged helicopter rotor blades from a crashed Sikorsky S-58, modified to allow blade-tip propulsion. The blades cost a bargain basement $50,000 – a significant saving since a new rotor blade would have been $1 m. Each rotor was powered by a hydrogen peroxide jet and was to be tested in a rock quarry. Approximately 5,000 lb of fuel provided up to five minutes of controlled flight.

Burt Rutan worked with Rotary Rocket to develop the ATV, with Scaled Composites responsible for the structural design, manufacturing and flight controls and systems. The ATV was flown on three test flights in 1999 by Marti Sarogul-Klijn, with Brian Binnie in the co-pilot's seat. On 28 July 1999, the pilots took it on three vertical flights for just under five minutes. It climbed to a maximum height of eight feet (2.4 m). Binnie, a man of dry wit, called the craft the Bat Cave, because the view was so restrictive the pilots had to rely on an altimeter to see if they had touched down. It was a hilarious project, if it hadn't been so dangerous. The whirling helicopter blade made the whole vehicle spin unless it was counteracted by jet thrust in the opposite direction.

Six weeks later, in September, the second flight was a little better, lasting two and a half minutes and reaching twenty feet. The third flight in October was the last. This time the ATV flew along the runway at Mojave airport to a distance of 4,300 ft (1,310 m) and reached 75 ft (23 m). But the whole craft was unstable. The funders weren't much impressed either – and a fourth flight was canned.

It all appeared to be a wasted journey for the Virgin guests. 'We both decided by late lunchtime that this didn't look like much of a prospect,' says Whitehorn. But, in the process, Richard caught up with Burt Rutan, whom he knew from past projects. Even Burt confided that he didn't think the Rotary project was going to be successful.

Sir Richard has always been fascinated by the flight of Voyager, flown by Burt's brother Dick, and Jeana Yeager (no relation to Chuck, as Dick frequently points out) in December 1986. Setting off from Edwards Air Force Base on 14 December, the plane – which looked like a flying catamaran – flew 26,678 miles around the Equator on a single tank of fuel. It remains an epic voyage – and a huge achievement for a bunch of Mojave amateurs and DIY plane-builders who gave up their free time to support Dick, Jeana and Burt, who designed the aircraft.

Burt was light years ahead of his time. Voyager's success would not have been possible without the technological developments. It was the largest all-composite airplane ever built, and the father of much of Scaled Composites' later work and of Virgin Galactic's SpaceShipTwo. The composite material was woven fibres – of glass, graphite or aramid – bonded with epoxies or other resins. Heated in an autoclave, the compound became immensely strong. The structures were far lighter than the pressed aluminium used for planes. Aramid is the generic name for Kevlar, now the staple for bulletproof vests, while Nomex is the trade name for the honeycombed fibre. Even to this day, Burt cannot bring himself to mention the trade names, owned by Dupont. He had approached them for funding for Voyager in 1986, but they demanded that the whole plane be built in Kevlar. When Burt pointed out it wasn't an appropriate material for a whole plane, they pulled the funding offer. Voyager was a superb achievement and President Ronald Reagan recognised this, honouring Burt, Jeana and Dick with the prestigious Presidential Citizens' Medal in December 1986.

After the damp squib that was the Rotary display, the Virgin team headed off to have hamburgers and fries in the Voyager restaurant at Mojave Airport. The eaterie, which boasts the slogan 'Aviation Spoken Here' on the blue awning outside, is a warm and informal place overlooking the runway. After his meal, Whitehorn pulled out his Biro and started sketching a V-winged bomber with four engines onto a paper serviette. 'I completely forgot about this,' he says now. 'But when we were sitting in the restaurant I sketched out this little drawing on a napkin. I have no memory of doing this, but when the Virgin Galactic project went live it was brought out and shown to me. I was flabbergasted.'

Jackie McQuillin had kept it as a keepsake in her drawer in Virgin management's London office in Campden Hill Road. It was a neat drawing with some numbers, which foresaw Virgin Galactic's interest in space. 'We were discussing the downside of the Rotary rocket and I said we could have a big aircraft like a B52 or a Boeing 747. And I drew a very rough B52 bomber with four jet engines. Then we have a little spacecraft – made from lighter composites – which would blast off into space with paying passengers.' Then he scribbled a tiny little rocket ship that actually looks similar to SpaceShipOne.

'I said, "We've got to get the cost per passenger down to the $100,000 range, with about $100 million of development costs, if this thing is to make any sense at all",' recalls Whitehorn.

'When I look at it now, it's uncanny. I even had a couple of little jet motors at the back. I presumed you'd have to fly it like that, but I didn't know anything about feathering or the types of rocket motor. I just said build a little rocket, sling it under a big airplane, fly it up and let it go.'

The humble napkin is often overlooked as a source of inspiration – indeed perhaps another tourist attraction could become the World Napkin Museum, because Burt Rutan sketched his Voyager designs on a stained serviette from the Mojave Inn, nearly twenty years previously.

But another course of events came into play. Alex Tai, who had undertaken a number of record-breaking flights with the explorer and adventurer Steve Fossett, made a request. He had also been a keen follower of the Voyager flights of Dick Rutan and Jeana Yeager, and now wanted to know if it was possible to do an around-the-world trip single-handed? Surely no human being could stay awake that long, and control the flight – after all, it had taken the Voyager pair nine days nonstop? But Alex saw a great plan being hatched.

Alex Tai is Virgin Galactic's Chief Operations Officer and he joined Virgin in 1995 as a pilot. He is likely to be the pilot of the inaugural SpaceShipTwo flight taking Sir Richard Branson and his family into space.

Tai has always been a flying fanatic. At the age of twelve, he joined the Air Cadets near his home at Bedmond, between Watford and Hemel Hempstead. When he moved on to Rickmansworth Grammar School, he was taken along to the 2nd Founders Squadron at Watford. 'It was there that I got the bug for aviation. I was given air experience and flights and did a gliding scholarship when I was sixteen. I did fairly well and they asked if I'd like to become a staff cadet.'

He flew solo at sixteen in a Venture TT motor glider, and

was then selected for an RAF scholarship, which paid for his private pilot's licence at seventeen.

'The Air Cadets really inspired me to go into the air force and I wanted to be a fast jet pilot.' Soon after completing his A levels, he was accepted on a permanent commission as a trainee pilot. He graduated from Cranwell and then went to RAF Church Fenton, Yorkshire, where he successfully completed the course and was streamed to fly fast jets.

But young Tai was about to learn a tough lesson. While he was talented and capable of hurtling a Jet Provost trainer around the sky, he was too young to appreciate the dangers.

'Basically, I was overconfident and, at the age of eighteen, flying fast jets in the air force I didn't think I could die. It was all too easy. This exhibited itself in me being dangerous at low level. I was told that I was going to end up as a steaming hole in the side of a hill. So the fast-jet option was taken away from me. I was very disappointed.'

Rather than stick with the RAF flying larger multi-engined aircraft that were slower, he left to pursue a commercial flying career. 'I did all sorts of things to build up my hours to get my commercial licence and then settled into a career flying executive aircraft, including Citations, Lear jets, through to the 125s.' Then, at a more mature 27, he applied to become an airline pilot, joining Air World, a Manchester charter company, flying as a second officer on the Airbus A320. He spent a season with Air World and then applied to the major airlines. He was offered a few jobs, Dragonair, Monarch and others being keen to hire this young pilot. But Virgin Atlantic secured his services and he joined in 1995.

It was one of those fortuitous moments in life that put him into the path of Sir Richard Branson. On one flight, the Virgin boss popped his head into the cockpit to say hello – something he does on every flight – to the flight-deck crew. He got chatting about weather conditions and launch sites for a possible ballooning mission in North Africa. Tai, who was sitting in the right-hand second-officer seat, suggested an airstrip he knew in Morocco. It turned out to be the ideal spot – and Branson's thank-you was to ask Tai to fly the chase plane.

Tai was made a Virgin Atlantic captain at the age of thirty and kept in touch. So it was Alex who pressed Sir Richard to meet Steve Fossett, who was keen to take on another record-breaking project. His proposition was the GlobalFlyer, a single-pilot, single-engine aircraft designed for nonstop global circumnavigation. The structure of the plane – like Rutan's Voyager – would be made entirely from composite and ultra-light material.

The germ of Steve Fossett's attempt to fly a solo, nonstop, non-refuelled circumnavigation of the world began in 1999 after a dinner at Barron Hilton's Flying M Ranch in Nevada, when Dick Rutan explained it was now possible for someone to smash the record he set with Jeana Yeager thirteen years earlier. Steve expressed his interest in attempting this with a newly developed composite aircraft and a few weeks later Dick made the introduction to Burt at the annual Experimental Aircraft Association gathering in Oshkosh, Wisconsin.

In his book *Chasing the Wind*, Steve Fossett recalls his own excitement and trepidation about the project.

> Before Burt and I got too far into the planning stage, it became evident that business at Scaled Composites, headed up by Burt Rutan, was going through some rocky times. The company that owned Scaled wanted to sell it. This posed a risk that, if I went ahead and invested the necessary funds to get this project rolling, Burt might be unhappy with the new owner and take his loyal engineers 'across the street' to form a new company.

He was concerned his contract would be left in the hands of the new company, so when Burt eventually organised a management buy-out, Fossett, a multimillionaire in his own right, became a shareholder, taking a 13 per cent stake.

The solo airplane project was code-named Capricorn – a reference to the qualifying distance of the flight around the Tropic of Capricorn. About a year after designs were drawn up and worked on, Steve invited Sir Richard Branson to join him sailing his yacht *PlayStation* in the Heineken Regatta in the Caribbean. The team came first – breaking a speed record along the way – and later that night Steve, after a celebratory meal, showed off Capricorn's futuristic twin-hulled design conjured up by Burt Rutan. Branson had already learned about the plans from Alex Tai, but seeing them now rolled out as a blueprint fired up the Virgin boss's imagination. He requested a full-scale meeting back in London with Will Whitehorn.

'Steve came to see Richard and me at Holland Park,' says Will Whitehorn. 'We sat down and talked about the environment and how we were becoming concerned about what was happening. We could see the merit of this record-breaking attempt but the real attraction was the plane and the technology – which would be an all-composite construction.'

Days earlier, Sir Richard Branson had been in conversation with Phil Conduit, then head of Boeing Commercial Planes, about putting more carbon composite into the commercial jets to make them lighter. Virgin Atlantic had several Boeing 747s on their transatlantic routes and were considering their next purchases. The choice was Boeing or Airbus, the European plane builder. Boeing weren't ready to make that leap at the time, and Conduit said there wouldn't be a Boeing composite plane for at least a decade.

'There was no incentive for the manufacturers at this time. They were very slow, although Airbus was moving much faster than Boeing in embracing composite material in aircraft construction. But the whole prospect of flying around the world, nonstop, on one tank of fuel was compelling. Imagine this plane using less fuel per hour than a four-wheel-drive American SUV. So GlobalFlyer was a showcase for us for the future,' said Whitehorn.

Virgin took time to bolt down the details but decided to support Fossett by becoming a major financial sponsor and using Virgin's public relations clout. Sir Richard Branson made the announcement at a press conference on 3 October 2003 at the Science Museum in London – the plane would become the Virgin Atlantic GlobalFlyer.

Burt already had a lot of information in his mind. The difference was that GlobalFlyer was to use a jet engine, rather than a single-engine propeller, and it was not as fuel-efficient. More fuel meant more weight. Burt had to lay down some tough parameters for an ultra-lightweight plane that could break the record. Hundreds of calculations were done on the blackboard and on the computers to look at fuel economy. The aircraft would be flying at up to 52,000 ft (17,000 m) and travelling between 19,000 and 25,000 miles (32,000 to 40,000 km) at speeds in excess of 250 knots (285 mph, 440 kph). It needed a reliable engine too – flying on a single engine over thousands of miles of ocean is not for the faint-hearted. The FJ44-3 ATW engine was being produced by Williams Engines and had a modest, but perfectly adequate, 2,300 lb of thrust.

Williams International built small jet engines used for light business jets such as the Cessna Citation, the Raytheon Premier and the Sino-Swearingen SJ30-2. The FJ44 was a simple two-spool turbofan, designed for low operating cost and reliability. More than 2,000 have been built, flying millions of hours, but the GlobalFlyer would have a modified, experimental version of this production engine. Rutan thought it was the best he could find because it gave a higher thrust-to-weight ratio and had the fuel economy the bid required. For the airframe, Rutan built a lightweight carbon-composite body with a Heath Robinson contraption of mechanically controlled flight controls using a push bar and crankshafts.

The Virgin GlobalFlyer – call sign V101 – aircraft was 38.7 ft long and 11.8 ft high, with a wingspan of 114 ft. Without fuel the craft weighed only 3,577 lb, but with its 17 fuel tanks filled up it was 7 times heavier with a gross weight of 22,066 lb. The pressurised cockpit allowed Steve Fossett to lie down, but it was very tight.

It was a project that consumed a lot of time and had Virgin Atlantic salivating, but eagle-eyed Alex Tai spotted something else. He was deeply intrigued. He phoned Whitehorn, catching him on his mobile phone in the middle of a Virgin Rail board meeting.

Tai, normally cool and collected, was unusually excitable.

'Will, Will. There's something you'll love in Burt's work-shop,' he said conspiratorially.

'What is it?' asked Whitehorn, sensing something news-worthy, before being given some dirty looks from the board. 'Er, can I call you back?'

Whitehorn switched off his phone and continued the meeting.

An hour later, his curiosity at fever pitch, he rang Tai back. 'So what is it?'

'Burt's got a spaceship.'

'What!'

'He's building a spacecraft and he's entering it for the X Prize.'

Whitehorn was hooked. 'So who is he building it for?'

'He can't tell me that,' said Tai.

'Have you asked him?'

'Will, do you think I'm daft?'

'No.'

'Well, Burt is keeping schtum.'

Whitehorn couldn't wait to get out to Scaled Composites at Mojave in California. But as Virgin's head of development it was several more days before he could find a good enough excuse to see how the plane was progressing. Then he was shown into another hangar in the factory where a tiny white object with rounded portholes was being finished.

'I remember Burt taking me in to see the spaceship with Alex. I saw it and I got straight away what Burt was trying to do. I thought, this is amazing.

'Everything was done. It was built but it had no motor in it (although Burt showed me the motor too).'

Burt repeated that client confidentiality was paramount and he could not tell Will who had paid for this tiny craft. Whitehorn made it clear to Rutan that Virgin would be very interested in developing the technology – and going commer-cial with it if it was successful.

Whitehorn went outside to phone Richard Branson and have a calming Marlboro. He was stunned, elated and laughed at the comedy of the situation.

'With the best will in the world, fuck the GlobalFlyer, he's building a spaceship!'

But the identity of the spaceship owner was causing all kinds of intrigue. Names were bandied about like confetti and then rejected. Around the Virgin management team, the most popular question was: 'Who is building this spacecraft?' No one had the answer.

Alex Tai, on time off from his day job flying Airbuses, nipped out regularly to Mojave in a small plane. He was keeping a watching brief on GlobalFlyer, but still Burt would not reveal his client.

'Can we speak to the owner? Will you let them know we're interested?' pleaded Alex.

But Burt kept his secret.

In 2003, a month or so later, Richard Branson and Will Whitehorn were in the office of *Popular Science* in Park Avenue in New York and met Eric Adams, its aviation and defence editor. 'I was talking about the GlobalFlyer while Richard was doing an interview on why we were doing the project,' says Whitehorn.

Branson explained that Steve was aiming for the longest flight ever – one of aviation's fifteen 'absolute' records certified by the governing body of aviation record, the Fédération Aéronautique Internationale (FAI) – and that the developments in composite technology were the only way to allow aviation to be sustainable over the coming decades.

Adams is an influential writer with an enviable range of industry contacts. After the formal interview was over, Will pitched an in-swinger to Adams to sound out a response.

'And you know about Burt's spaceship?'

'Sure, I know all about the spaceship,' said Adams.

'Well, we'd love to develop it but we don't know who's building it,' replied Whitehorn.

'Well, I know, it's Paul Allen.'

The editorial team at *Popular Science* had been exchanging emails with Burt and lots of the technical people for the preparation of a major feature article and by accident the trail led back to Vulcan in Seattle, which was Paul Allen's company.

'Eric told us that,' says Whitehorn, 'so Alex then spoke to Burt and said, "Is it Paul Allen?"'

'Burt replied with a grin, "I can't tell you but I will contact the owner."'

By Easter, Burt reluctantly confirmed to the inquisitive aerospace industry press that a secret space project was up and running. But still he was unable and unwilling to reveal the sponsor's name. On 18 April 2003, Scaled Composites unveiled the existence of its space programme, saying that it had been going on for two years – the airborne launch plane was called White Knight, while the capsule was to be known as SpaceShipOne.

At a ceremony at Mojave, the actor Cliff Robertson introduced Burt Rutan, who explained the history and the components of the programme, while Dr Maxim Faget, designer of the early NASA space programme from the Mercury through to Apollo, Erik Lindbergh, the grandson of Charles Lindbergh, and Dennis Tito turned up to watch.

Several of Wernher von Braun's former German colleagues – now into their mid-80s – were also at the launch. They had worked at Huntsville – renamed Hunsville – and now believed that Rutan was a genius and the living reincarnation of their former boss.

But the identity of Burt's secret customer remained officially unknown.

9. THE RECLUSIVE BILLIONAIRE

P aul Allen is a sci-fi buff. The founder of Microsoft loved every episode of *Star Trek* with William Shatner. That's why he named his Vulcan business after the planetary homeland of his favourite character, Spock.

The imperturbable first officer of the starship *Enterprise* – played by Leonard Nimoy – was born in 2230 on the planet Vulcan to a Vulcan father and human mother. Like all Vulcans, he has pointed ears and green blood as well as superhuman strength and an amazing intellect. He is also a vegetarian. He is well versed in art, history and music and he won the second prize in the All-Vulcan music competition, playing the harp. Paul Allen shares a huge affinity with Spock. And Spock would have admired Burt Rutan.

In 2001, Allen approached Rutan and asked Scaled Composites to fulfil his vision of flying the world's first privately developed, reusable space vehicle. He was prepared to put up $30 m for the research and development of the SS1 system – his goal was to win the $10 m Ansari X Prize. But one of the conditions was that it was to be top secret. But once news of the craft reached the science and aerospace media in the US, there was huge speculation about who was putting up this kind of seed-corn money.

On 17 December 2003, Paul Allen finally confirmed what had become an open secret in informed aviation circles for over nine months – he was the sugar daddy behind the innovative SpaceShipOne project. The labour and birth were worth the wait.

This was a sparkling day for Allen as SpaceShipOne broke through the sound barrier during its first manned test flight. 'For me, being able to watch today's successful test flight in person was really an overwhelming and awe-inspiring experience. I'm so proud to be able to support the work of Burt Rutan and his pioneering team at Scaled Composites,' he said.

It might have been another genuine milestone in the history of flight but it was a rough day for the test pilot. Brian Binnie was one of a trio of test pilots on the Scaled Composite's suborbital programme, along with Mike Melvill and Pete Siebold. Binnie flew SpaceShipOne's first flight that day.

'The test card was ambitious; because of the prodigiously rapid acceleration of the partially fuelled ship, the first powered flight was also to be the first supersonic one,' according to *Flying* magazine. 'The transonic flight regime, where the speed of the aircraft is subsonic but the flow over some portions of it is supersonic, is treacherous: it is there, rather than in fully supersonic flight, that the worst aerodynamic surprises are likely to occur.'

There were no problems in this area. The 15-second burn produced a speed of Mach 1.2 and a maximum altitude of almost 68,000 ft. But the landing was heavy and the left main landing gear failed. 'There were technical reasons for the hard landing – viscous control dampers installed to protect against flutter had stiffened up in the cold at high altitude – but Binnie had to live down a lot of sarcastic references to flareless carrier landings. He despaired of being allowed to fly the ship again,' said the *Flying* article, in January 2005.

This was of little consequence to Paul Allen, though. In his eyes it was an astounding achievement.

'Paul shares our energy and passion for not only supporting one-of-a-kind research, but also a vision of how this kind of space programme can shape the future and inspire people around the world. Today's milestone and the SpaceShipOne project would never have been possible without Paul's tremendous support,' said Burt Rutan after the flight.

On 21 June 2004, when Mike Melvill flew SS1 above 100 km altitude, Allen was elated. Here was a significant breakthrough, dispelling the myth once and for all that manned space flight was the sole domain of huge government programmes.

Melvill has been a close friend and business partner of Burt Rutan and his brother Dick since the late 1970s. He had been a fascinated DIY hobbiest who used one of Burt's designs to make his own home-built plane. But he was also a skilled pilot and went to live and work with the RAF – the Rutan Aircraft Factory, the forerunner to Scaled Composites.

Meanwhile Sir Richard and Paul Allen had been in talks too. There was a growing belief by Allen and his advisers that Virgin's international clout could really make the difference in building a true commercial space operation. So Alex Tai and Will Whitehorn, along with John Peachey, Virgin's investment director and a Galactic board member, flew to Seattle to meet Allen's advisers. Peachey was the moneyman, the reality check to keep Whitehorn's boundless enthusiasm under control. 'He was there to make sure I didn't make any rash promises!' said Whitehorn.

The first deal didn't work for Allen's company. 'We couldn't get the deal to close. I think for Vulcan it was funny that they had never done a deal where people were paying them money for something. They were quite nervous about that,' recalls Whitehorn.

But discussions resumed again shortly after – and this time things progressed well. It was extremely tight if they were going to get the deal done before the Ansari X Prize. A few days earlier, Alex Tai had been sitting at his kitchen table in

Notting Hill sketching out his wish list and outlining what needed to be done. One of the jobs was to find a logo. He phoned up a local London company who thought it was a joke and a typical Virgin publicity stunt. But they came up with an excellent stopgap sci-fi logo in red and grey which passed muster.

Virgin eventually negotiated with Paul Allen to buy the rights to use his technology.

'It was a push. We signed the head of terms and the final documents were signed over a Saturday at my home in Wadhurst, in East Sussex, and faxed back to the US. This was three weeks before the X Prize. It was a terrific deal for us because the Virgin Galactic branding would now be on Paul Allen's SpaceShipOne during the prize in the Mojave. This would give us worldwide exposure – and it would deliver a message that we were now a serious player,' adds Whitehorn.

Once the agreement was made, Virgin's tenacious press team of Jackie McQuillan and Terri Razzell hit the phones to get the announcement out in time. It was going to be a big stretch. But they managed to entice every Fleet Street newspaper and a battery of television channels to cover the event. Two days later, on 27 September 2004, Sir Richard Branson was on the platform at the Royal Aeronautical Society in London with Burt Rutan to announce the launch of Virgin Galactic, the world's first commercial space tourism operator.

'We hope to create thousands of astronauts over the next few years and bring alive their dream of seeing the majestic beauty of our planet from above, the stars in all their glory and the amazing sensation of weightlessness. The development will also allow every country in the world to have their own astronauts, rather than a privileged few,' Sir Richard told the gathering. He revealed that the first flights would be in Virgin SpaceShip *Enterprise*, a reference which doffed its cap to *Star Trek*.

This was the birth of a new era.

Sir Richard Branson told the press pack that 'following the successful conclusion of these negotiations, we signed a $21.5 million deal for the use of that technology and developed a

$100 million investment plan to build up to five spaceships at Mr Rutan's factory in Mojave'.

Paul Allen, who was not there in person, said, 'I backed the development of SpaceShipOne because I saw this as a great opportunity to demonstrate that space exploration could some day be within the reach of private citizens. Today's deal with Virgin represents the next stage in the evolution of the SpaceShipOne concept, and will likely be the first of a number of deals that will utilise the technology developed during the creation.'

Next morning's tabloids didn't give the new venture an easy baptism. As Branson was unveiling the space deal, his Virgin Rail's advanced Pendolino tilting train service from Glasgow to London's Euston station was snared by technical faults. The famous Royal Scot service was hit by a wheel problem and forced to slow down from 110 mph to 50 mph. Commuters were late and expressed their anger. It was only teething problems for a new service, but the mid-market *Daily Mail* couldn't resist a brilliantly sardonic headline: 'Euston, we have a problem ... as Branson unveils his Virgin spaceship, his new trains hit trouble.'

Tilting trains aside, Branson, Rutan and Allen would be a dream team to ignite worldwide interest.

But how and why did one of the world's richest men want to make his mark on space? The story goes that Paul Allen saw an advert for a cut-price, DIY computer, the Altair 8800, made by MITS of Albuquerque, New Mexico. He went off to show it to his close friend Bill Gates – they had been at prep school together. Gates was then a student at Harvard.

They bought the Altair and then called the president of MITS claiming that they had written a version of BASIC, a program language, for the Altair computer. In fact, Gates and Allen had just started working on their version of BASIC and it took them six weeks of sleepless days and nights to finish it. Their version was GW-Basic (or Gee Whiz). BASIC had been created in 1964 by John Kemeny and Thomas Kurtz, two Dartmouth maths professors, and they allowed it to be 'open access' so it fell into the public domain. GW-Basic was an instant success. In 1976, Allen and Gates, then only 19 years old, founded Microsoft. Allen served as the company's executive vice president of research and new product development, the company's senior technology post, until 1983. Allen left after being diagnosed with Hodgkin's disease. He was told there was no chance of survival from the lymphoma, but after a two-year battle he beat the disease.

Paul Allen is rich beyond his wildest dream. Even he is unlikely to know how much he is really worth. He has invested in a range of companies exploring the potential of digital communications. His business strategy spans the areas of technology, new media, biotechnology, telecommunications and entertainment. He is also a partner in the entertainment studio DreamWorks.

Vulcan Inc. of Seattle and Charter Communications of St Louis, America's fourth biggest cable provider, are his main businesses. But Allen has sporting interests too, as owner of the Portland Trail Blazers National Basketball team and the Seattle Seahawks National Football League franchise. Allen gives a lot back too. He is one of the world's leading philanthropists through the six Paul G Allen Charitable Foundations, which support arts, health and human services, medical research, and forests in the Pacific Northwest. He's the founder of Experience Music Project, Seattle's interactive music museum, and, something closer to his heart, the Experience Science Fiction Museum, a master collection for sci-fi buffs – and those interested in space. It opened in 2004 and has attracted millions of visitors.

Included in the exhibits are the works of the giants of the SF genre: Isaac Asimov, H G Wells, George Lucas, Arthur C Clarke, Ray Bradbury, Gene Rodenberry, James Cameron and Steven Spielberg. From Mary Shelley's *Frankenstein*, first published in 1818, and *The Birthmark* by Nathaniel Hawthorne, to H G Wells' *War of the Worlds* and George Orwell's *Nineteen Eighty-Four*, to the cartoon series *The Jetsons*, the BBC's *Doctor Who* and the Warner Bros film *The Matrix Reloaded*, Paul Allen argues that science fiction remains one of the most compelling, popular and intriguing genres of human expression.

'Science fiction has always been a vehicle for entertainment, but more importantly it's a genre that is forward-looking by nature, expanding people's views of science, technology and the future – and their exciting possibilities,' he says.

The Seattle experiences are designed to show visitors a variety of science fiction stories, concepts and social commentary. But the real drama was about to happen in the purple-and-black ceiling high above the Mojave Desert.

Peter Diamandis felt relieved that the event was now happening. 'There was a sense that the faith people had put in me over the past decade had not been misplaced. So many had backed me, given me their money and supported the idea – so getting the X Prize won was not an option, it was a mandate. Given that we funded the prize through a hole-in-one insurance policy that was set to expire on 1 January 2005 ... it was won with only two months and three weeks to spare. That was a bit stressful,' he says.

The Ansari X Prize itself had 29 entrants, but only 3 serious contenders – and only one of these had serious funding – Paul Allen's SpaceShipOne. Flying from Mojave Airport's Civilian Aerospace Test Center on 29 September 2004 and then repeating the feat on 4 October was the breakthrough.

Mike Melvill, a long-time friend, associate and the right-hand man of Burt Rutan, was the courageous pilot who put all his faith in the innovative rocket design and the unique shuttlecock-feathering re-entry mechanism. It was a shaky ride, which required brilliant skills from the pilot. But some observers had the audacity to question Burt's flightpath. Rutan was pressed to give some explanation for the extreme rolling motions in the first X Prize flight. He was angered by some of the gossip on the plethora of websites following the flight.

'While the first roll occurred at a high true speed, about 2.7 Mach, the aerodynamic loads were quite low and were decreasing rapidly, so the ship never saw any significant structural stresses. The reason that there were so many rolls was because shortly after they started, Mike was approaching the extremities of the atmosphere,' said Burt on his Scaled website.

What the space buffs witnessed was something very new and Burt had to set the record straight. 'Nearly all of the twenty-nine rolls that followed the initial departure were basically at near-zero-g, thus they were a continuous rolling motion without aerodynamic damping, rather than the airplane-like aerodynamic rolls seen by an aerobatic airplane.'

In other words, this was more like a space flight than an airplane flight. So Mike Melvill could not smooth out the flight with his aerodynamic flight controls. Rutan explained that Melvill decided to wait until he feathered the Space-ShipOne's tail in space, before using the control system thrusters to damp the roll rate. 'When he finally started to damp the rates he did so successfully and promptly,' said Burt.

This was stable attitude, without the craft wavering, before the SpaceShipOne reached its apogee of 337,600 feet, or 103 km. This was space. And it gave Mike time to relax, note his peak altitude, and then pick up a digital high-resolution camera and take some photos out the windows. According to Burt, there was no problem.

'While we did not plan the rolls, we did get valuable engineering data on how well our system works in space to damp high angular rates. We also got a further evaluation of our re-entry capability, under challenging test conditions,' explained Burt.

SpaceShipOne righted itself quickly and accurately without pilot input as it fell back into the atmosphere. 'No other winged, horizontal-landing spaceship – not even the X-15 or the space shuttle – has this capability,' said Burt proudly.

But one other issue was bugging the life out of Rutan. 'Some publications have stated that Mike defied a request to shut down the motor and let it run a few more seconds in order to reach 100 km altitude. This is not true. While a mission control aerodynamist did discuss a possible abort a few seconds earlier, Mike immediately shut down the motor on the first advisory call over the radio. Mike himself was monitoring the apogee predictor during the initial rolls and was in the process of going for the thrust as he heard the advisory call.'

Over the next few days at Mojave Airport there was a lot of technical talk over countless cups of thick, black coffee. 'Mike Melvill flew the first X Prize flight on the Wednesday. Then we spent most of Thursday investigating the likely cause of the directional departure he had experienced,' recalls Brian Binnie.

'Burt, convinced we were on the right track, announced we would fly again in four days – or the next Monday. I was informed late Thursday night that I'd be the pilot,' he says.

There was a colossal amount of interest from around the world. There was a media frenzy about this tiny little craft that looked like an overgrown shuttlecock. And Burt's announcement gave thousands the chance to head out to Mojave. The press had tried to paint a two-horse competition between Scaled Composites and the Canadian company, the Da Vinci Project, with their DreamSpace craft. While Brian Feeney, Da Vinci's president, might have been a wizard designer with his tantalisingly futuristic craft straight out of a *Star Wars* movie, he couldn't get the backing for his rocket ready in time. In reality, there was only one bucking bronco left in this rodeo. And Binnie was about to ride it.

Peter Diamandis was in clover. 'The early rules were created by myself. I ultimately put together an outside advisory board of aerospace experts to discuss and help refine the rules. We had a lot of debate about whether it should be a hundred miles or a hundred kilometres . . . one seat versus three seats. The greatest compliment Burt Rutan has ever paid to me was that the rules were "spot on" and had held true for eight years,' he says.

On 4 October 2004, SpaceShipOne with Brian Binnie at the controls was launched from its mother ship and soared into suborbit, reaching 367,442 ft above the Earth for the second time within fourteen days. For Binnie, it was a flight and a day to remember for the rest of his life. He had become an astronaut. After the flight, he was informed by Rick Searfoss, the X Prize judge and a former shuttle commander, that he was astronaut number 434.

'I was humbled and delighted,' says an emotional Brian Binnie. 'But certainly there are different colours of astronauts.

Currently there is no distinction in the vocabulary between Neil Armstrong as an astronaut and someone who has just completed the initial ground training at NASA and who may not actually ever get to space if the shuttle fleet is grounded by 2010 as the new administrator suggests. Neither is there any distinction between the pilot types and the mission specialists.'

But Binnie is only the third astronaut to have actually *flown by themself* into space. Joe Walker, the record-breaking X-15 pilot, was first and Mike Melvill was second. The person with the chequebook who had made science fiction a reality was Paul G Allen.

10. BURT: A LEGEND OF MOJAVE

It's a straight drive of eighty miles from Los Angeles up Highway 405 North then Highway 5 North, to Burt Rutan's inner sanctum. Beyond the city smog, the land becomes a burned-ochre stage set for a hundred cowboy films, stretching out ahead against a cloudless azure canopy. On the parched slopes ahead row upon row of wind generators – like tiny needles pricked into the ground – spin languidly in the burning afternoon. A large concrete sign, framed with local rocks, welcomes you to Kern County and 'The Home of the Voyager' when you hit the dusty town of Mojave. Mojave was a railway town long before the biplane arrived. On the left are the tracks of the Atchison, Topeka and Sante Fe Railroad, with large diesel engines delivering a mile-long chain of lumbering freight.

As the airport perimeter sign comes into view the road curves past a row of nondescript sheds, then the logo of Scaled Composites comes into view. For an aircraft manufacturer, it is an anonymous kind of place. The windowless complex is right on the airport front, and through a large wire fence you can see some of Burt's planes. But the content of the sheds remains top secret.

Visitors can stand in the glass front porch, but won't get through the security gate unless they have high-level, written clearance. Inside, Burt Rutan's office is a cosy place, his desk piled with papers, books and several computer screens, each plastered with pink and white Post-it notes with key numbers such as the local taxi firm. It's also a living museum to his genius. On the office wall is an enlarged black-and-white picture of a young Burt playing with his model airplanes, and then working as a civilian in the air force.

In the boardroom there are more pictures. It's a fantastic gallery of aviation. There are dozens of plane designs, some that have never left the draughtsman's pad, but others in various stages of manufacture, production and in flight. On the shelves and sideboards are loose-leaf folders, manuals and models of a plethora of experimental aircraft, from the Boeing X-37 to the X-15, along with the Vari-Viggen, Vari Eze, Proteus and the Boomerang, all created by Burt, and now the model of SpaceShipOne. All along the corridors are the front covers of magazines from *Popular Science, Flying, Aviation Week* and *Flight International,* all with Rutan designs on the cover. It's a veritable Aladdin's cave for any plane-spotting schoolboy.

Much of Burt Rutan's earliest thinking was a legacy of airplane modelling, which enjoyed enormous popularity when radio-controlled mini fliers arrived in the late 1950s. Burt was a teenage expert. Indeed, the trailing edge of the Voyager wing was originally made of light balsa wood and fabric. Burt, born in Oregon, grew up in the 1950s in the small town of Dinuba, in California's Central Valley, near Fresno. After his brother went off and joined the air force, Burt was persuaded by his parents to go to college at California Polytechnic in San Luis Obispo. Here he began work on his first plane, the Vari-Viggen.

According to Dick Rutan, 'All of Burt's airplane designs have been in a sense an extension of his fascination with building model airplanes. The Vari-Viggen, his first airplane, was like a giant model, built mostly of plywood instead of balsa. Its handling was intended to provide in a tiny airplane an accurate scale rendition of the feel and handling of much larger and faster jet fighters.'

After college Burt went to work at the Edwards Air Force Base as a civilian test engineer and was placed on the F-4 Phantom programme. The Phantom was a standard fighter-bomber during the Vietnam War era, but most pilots loathed it because it often went into unprovoked spins. Burt was assigned to help sort this out. The only way to do this was practically by forcing the plane into a spin, so Burt flew in the back seat. On one flight test the spins were so severe that both Burt and the pilot, Jerry Gentry, were forced to bale out. But Burt solved the issue. On the basis of his test flights, he prepared a manual for pilots on how to get out of a spin and became known as the man who saved the F-4 programme.

He was always keen on his model planes and, armed with a $10,000 loan from his father, he built the Vari Eze, which made his name in the annual Experimental Aircraft Association Fly-In at Oshkosh in 1975. It was made of fibreglass composite – and this became an aviation hobbyist route to the skies. The build-your-own plans were sold for $50 with 800 built and flown.

Burt Rutan was in demand. He was a consultant on new technology for Lockheed and other mainstream aircraft builders. Then, on spec, he sent NASA one of his designs. The engineers made encouraging noises. Here was a curious 'skew winged' airplane – the wings like a giant open pair of scissors – which was intended to test ways of reducing transonic drag and could be built for a modest £250,000. At the time, NASA's prized engineers were working on space shuttle designs costing tens of millions of dollars. Burt's design was rejected on the grounds that if it didn't cost a lot, it obviously wasn't good enough.

Undeterred, Burt was developing and refining another of his trademarks – putting the wings at the *front* of the plane. It had cropped up in the pioneering aero designs and was a feature favoured by the Wright Brothers, but had gone out of fashion before the First World War. The front wing was called a 'canard'. It was a life-saving device because the early fliers found the most dangerous feature of flying, both gliding and powered, was when a plane stalled. The canard – and the nose of the plane – dips if the plane is about to stall and this increases speed and restores the lift. Burt Rutan could see that the canard was an automatic pitch regulator, but it had been abandoned when propellers were put on the front of planes. And Burt's aircraft, from the first Vari-Viggens and Vari Ezes, had the props on the back to simplify the air flow. So front wings became a Rutan signature.

And here Burt paid tribute to the Swedish Saab Viggen jet fighter that reintroduced the canard wing in 1962 to aid manoeuvrability and takeoff from short stretches of motorway. The plane was designed to be stored in bomb-proof shelters, then rolled out to take off from runways as short as 1,500 ft in the event of a war. He named his plane the Vari-Viggen in homage to the Swedes. Burt was searching for the kind of jet-fighter handling in a home-built plane and he achieved this with the Vari Eze and the Long EZ.

Followers of Burt Rutan would buy his designs and turn out their own versions of his plane. They are the weekend DIY mavericks who still swap tips, share technical specifications and trade flying stories through dozens of website links. His planes instil a loyal kinship – on his fortieth birthday they threw a fly-in party for him and 45 owners brought their planes to Mojave. As a thank you, Dick, Burt and Mike Melvill flew in an aerobatic display.

On 27 July 2005, at Oshkosh, Wisconsin, Richard Branson and Burt Rutan announced the signing of an agreement to form a new business. It was agreed that the new company would own all the designs of SpaceShipTwo and the White Knight Two launch systems that were being developed at Scaled Composites. The newly formed business, the Spaceship Company, would be jointly owned by Virgin and Scaled. Alex Tai would be its senior manager. Burt's company would undertake all the research, development, testing and certification of the two crafts, with Burt heading up the technical development team. Virgin paid $21.5 m for an exclusive licence for the core design and technology and another $50 m was going to Scaled Composites to build five passenger spacecrafts. Another $70 m was earmarked for a Virgin base in New Mexico.

Burt Rutan is now a fresh-faced 65-year-old with strong, white teeth. He has clear green eyes, and thinning grey hair and, with his distinctive mutton-chop sideburns, he looks like an old Wild West gunslinger. In relaxed company, he is a raconteur and scurrilous joke-teller, but he has no time for small talk with strangers. And his media interviews are brief, technical and to the point.

Burt's factory is out of bounds for casual drop-ins, so our interview took place in a favourite haunt – the Voyager restaurant at Mojave airport. It was the evening of a special gathering – the twentieth anniversary reunion of the team who helped fly the Voyager around the world – and the end of a weekend of celebration, misty-eyed nostalgia and sharing a few drinks too many with friends who have all aged gracefully.

Spurred on by the gregarious host for the evening, Kelly Hall, Dick and Burt then regaled the group with some aviation tales – and a few near-the-knuckle jokes. But this was a home crowd: intimate, knowledgeable and safe.

Then the mood became more wistful as Dick reflected on his missed opportunities. 'Sometimes after the Voyager flight maybe I didn't do too well with contracting and lawyers and stuff, and any time I was feeling sorry for myself, I could walk in to the National Aerospace Museum in our nation's capital. When I walk in the door the very first aeroplane that I see is that Voyager. And I can walk up and stand underneath it and look up and say to myself – or out loud – I built that son-of-a-bitch and I flew her around the world,' he concluded, to a huge ovation.

Then Burt Rutan revealed another little titbit kept secret for twenty years – that this bunch of amateurs and volunteers in the Mojave Desert was being watched. 'One thing we didn't talk about at the time. It was kind of a spooky thing. About day one or two, we got a call from someone who said: "Listen, I can't tell you who I am, but we photograph your airplane all the time and if you ever lose track of it, call this number," and he hung up the phone. So we put the number in a drawer and never had to look for it.' But this gave Burt confidence that although the team had only a crummy, phone-patched VHF communications signal for locating the plane, that someone else was able to give them help if the Voyager got into trouble over the wide oceans. 'We knew, but not everybody knew, because we didn't tell you that we could find out exactly where it was at any time. That was kinda cool.'

After the photographs were taken and the backs slapped, and the dishes washed up in the kitchen, it was time to have a discussion with Dick about more recent events.

I wanted to ask him about the feathering process. Peter Garrison, a contributing editor to *Flying* magazine, has his theories about Rutan's success with SpaceShipOne. One was that 'despite his association with advanced technologies', Rutan prefers 'simple, straight-forward solutions'.

The other was the 'part-ingenious, part-fortuitous invention he calls "feather" – the folding of the airplane into the shape of a partially opened jackknife – that solved the problem of dissipating energy, while maintaining stability during re-entry'. The feathering process is the essential element of the whole system; without it nothing else would matter, reckoned Garrison.

So I asked Burt how he invented the revolutionary feathering process that made his spaceship so unique.

'Well, it is best to talk about the requirement first. It was a given that a winged vehicle that re-enters the atmosphere, that is able to glide in and land, rather than one that is recovered by parachute, on re-entry has very narrow constraints in terms of the angle in which it goes into the atmosphere and the attitude,' says Rutan.

He then recounts his deep and prolonged interest in the X-15. 'That plane flew twice above a hundred kilometres and it then had to re-enter the atmosphere at forty degrees flight-path angle, and essentially a forty-degree angle of attack – and a very small amount of sideslip.'

Sideslip is the ability of the spaceship to make sure it is not wobbling from side to side. This requires an excellent set of skills from the pilot. The angle of attack is a critical factor between the aircraft's longitudinal axis and the airflow.

'The space shuttle re-enters the atmosphere extremely shallow but at a high angle of attack and also [within] very narrow limits of sideslip. The X-15 flew for 199 flights and there was only one fatal accident and that was caused by the pilot not carefully controlling the attitude there in the atmospheric re-entry. And that haunted me from a standpoint – because I always thought that this was the biggest risk. You can improve the performance of what they did – in terms of flying the public out of the atmosphere – but meeting the precise control on atmospheric re-entry was, I thought, one of the generic risks,' he adds.

'So, initially, because of this, I decided that my suborbital spaceship would not be a winged vehicle that would land at an airport. And I resigned myself to the fact that we were going to have to accept the limited reliability and operational utility that you get with a parachute. It's a big cost-and-safety issue and also you can't operate in a lot of wind.'

But this wasn't really a satisfactory solution for a plane-builder who had spent a lifetime examining this very issue. Rutan wasn't going to give up – and besides, an old-technology solution was never going to secure the Ansari X Prize.

'So, I gave myself a challenge and put in front of myself a requirement. You must understand, this is a very difficult requirement for an aeroplane that is controlled manually because it can't have a fly-by-wire flight-control system – because that is too expensive.'

A fly-by-wire system is now the state of the art in most commercial airplanes. Originally conceived by Airbus Industries, it is a computer system that would tell the pilot how to control the aircraft more effectively and efficiently. For many years the rival designers at Boeing resisted a full fly-by-wire system, arguing that it took complete control out of the hands of the experienced pilot. But now most modern commercial airliners have fly-by-wire computerised control. This gives efficiencies in terms of fuel economy and takeoff and landing.

'The spaceship has to be hand flown, like a light airplane, and yet it has to go now through its atmospheric deceleration, supersonic and then transonic speeds. Getting something that would re-enter the atmosphere at a high enough drag and . . . to be controlled precisely, I felt was a showstopper.'

Burt then spent many months pondering this basic conundrum – which was in many cases the Holy Grail of space flight. 'I came up with a number of configurations that we checked with computational fluid dynamics (CFD).'

But Burt was keen to dispel one myth that has taken wings. Burt and the team at Scaled Composites made some paper airplanes, using pieces of A4 paper to examine various shapes and sizes, and launched them off the old four-storey control tower at Mojave Airport. 'Just for some fun we threw some models off the tower. That wasn't really a development. It was a nice angle for some reporters though. This was a stunt that we did for kicks. If you look at some documentaries you get the impression that was how we developed it. That's not true at all.'

Burt found that all his early ideas were not working. 'They were all tested and analysed at subsonic speeds; when I got around to doing the supersonic CFD, I found out that none of them worked.'

The unique movement of SpaceShipOne, where it raises a large portion of the wings and the tail outboard at an angle up to 90 degrees, was the first configuration that he stumbled upon. It looked promising. Scaled Composites made some small-scale models and ran the computer numbers.

'This was the first configuration that could be thrown into the atmosphere at any attitude and it would straighten itself out and the pilot didn't have to control it. It can be thrown at the atmosphere at any flight-path angle – and you don't hurt anybody. So I felt that was the most important breakthrough that could lead us to go directly to a system that would fly the public,' he says. 'I am not saying that this is the only configuration that will work – it is just the first one that we came into.'

SpaceShipTwo will be an adaptation of many of these revolutionary features, but Burt was keen to keep the next phase of the technology close to his chest until all of the testing is done. 'I can't talk about that SS2. I don't want our competition to know what we are doing – and I certainly don't want them to know the details of how I am solving the problems. And I don't want them to know the schedules.'

While the X Prize was undoubtedly a high-water mark for Burt and Scaled Composites, it is in the recent past – and now historic. It was called Tier One – which meant going suborbital.

But SpaceShipOne's control system, designed and built by Scaled Composites, and called the Tier One Navigation Unit (TONU), gives us a hint about the future. Every bit of avionics, instrumentation and the monitoring systems are in this essential black box. TONU is a digitally generated instrument panel that is the pilot's primary focus during much of the flight. And because the space pilot flies this by hand, it tells him or her everything they need to know – and the exact state of the vehicle. In the centre of the display is the horizon indicator that shows the craft's attitude. A V-shaped bar indicates the nose and wing attitude. While the computer provides the information, it is up to the pilot to fly the spaceship. *Flight Journal*'s editor-in-chief Budd Davisson, who underwent simulation training on SS1, said, 'My everlasting impression of the entire Tier One programme and the SS1 will be that it's brilliantly simple.'

The unveiling of SpaceShipTwo takes Scaled to the next level. The new generation of Tier One Navigation Unit 2.0 (TONU2) is more sophisticated to reflect the size and the passengers on board, but has exactly the same characteristics for flying – keeping it as simple as possible.

But there is now Tier Two and Tier Three to think about. Tier Two is orbital, while the third step is interplanetary.

Rutan hopes to accomplish all of these challenges. At an age when most mortals are thinking of retirement, Rutan feels there is no time to waste in making the next breakthrough.

11. THE FLIGHT OF YOUR LIFE

I n a world bereft of real heroes, Brian Binnie is the genuine article. Cool, calm and charming, he is the kind of person you would pick to have close by if you faced a crisis. And he's exactly the kind of pilot who will soothe the jagged nerves of those first Virgin space tourists. He's worth meeting in his natural habitat.

In an Aberdeen hotel room, Brian Binnie is consulting some charts in a ring-pull folder. He is marking some co-ordinates with a red pen while sipping a tin of ice-cold Dr Pepper. Binnie is a programme business manager and a test pilot at Scaled Composites. Even on a break to Scotland, he is working on some pressing figures.

'Hi, how are ya?' he asks. 'Come on and sit down and we can have a chat.'

Binnie has thin dark hair and a muscular face. He looks fit and lean for a man in his mid-fifties. He still retains the bearing and stature of a navy jock, although it is seventeen years since he flew on combat missions. He is cautious about what he says – and respectful of his chosen profession: test pilot. Often there is a fine line between success and disaster. Cautious optimism is the default state of mind for people who push planes to their extremes, but memories of failure are still etched in the mind of many at Scaled Composites. Chuck Coleman, a friend and colleague, crashed his plane weeks earlier and was still seriously injured. Chuck had been practising an aerobatic routine in an Extra 300 for an upcoming air show during a flight from Mojave Airport, and his recovery is still a topic of concern among the close-knit community at Scaled Composites.

But Binnie has been interested in planes since he was a boy back in rainy Scotland – a world away from the arid Californian desert. Binnie is back in Scotland, not to play golf – although he has a passion for the links – but to be given an honorary degree from the university where his father once taught.

When he was five, Binnie remembers his father, Bill, taking him out into the wide-open space of a golf course outside Aberdeen and flying model airplanes. His mother, Catherine, and father came from Camelon, near Falkirk, in the industrial central belt of Scotland, but both headed north in 1940 to attend Aberdeen University.

Bill went on to teach physics at the university and instilled in his son an interest in practical science. When he was offered a position as a crystallographer in Canada he grabbed the chance. He worked in Montreal for a while before being offered a job at Purdue University, Indiana, where Brian and his younger sister, Sheila, were born.

Bill brought his young family back home to Scotland where Brian was educated in the rudiments of flight. 'Our favourite model was a folding-wing glider you launched with a slingshot and, when the airplane slowed down near apogee, the wings would unfold and start a lazy descent back to ground,' he recalls.

This wasn't quite SpaceShipTwo, but folding wings are still a part of the Binnie life story. Binnie has some of the *Braveheart* Scottish spirit in his blood, although he is a proud American who has served his nation.

He told the *Sunday Post*, Scotland's old-fashioned Sunday newspaper: 'While I wasn't born in Scotland that's not my fault! All my relatives are Scottish and I grew up and lived there and spent my youth playing in the wood and hills and, of course, football and golf.

'When I came back to the States, kids would surround me in high school and make me talk just to hear my Scottish accent.'

Will Whitehorn, himself a native-born Scot, was surprised when he read that Brian was the first Scotsman in space. 'I didn't find out he was Scottish until after the X Prize flights.'

The Virgin Galactic chief learned of Binnie's Scottish roots while reading the *Daily Record*, one of Scotland's biggest selling tabloid newspapers. Whitehorn had always dreamed of being the first Scot in space, although it has to be said that Neil Armstrong was of Scottish border stock. But there were no hard feelings as Whitehorn informed his old university of Binnie's connection; in July 2006, Binnie – with his mother and father in attendance – was given an honorary doctorate. 'It was a very proud moment for me. I was touched to receive such an honour from my father's university.'

Binnie took a Bachelor of Science in aerospace engineering and Master of Science in fluid mechanics and thermodynamics from Brown University, Rhode Island, and then another Master of Science in aeronautical engineering from Princeton University. He then became a navy flier, who graduated from the test pilot school at Patuxent River, Maryland. With 21 years flight-test experience, including 20 years in the US Navy, he has logged over 4,600 hours of flight time in 59 different aircraft. He is a licensed Airline Transport Pilot and a member of the Society of Experimental Test Pilots, an illustrious body for those with the Right Stuff.

During his two decades in the US Navy, he flew F18s from aircraft carriers and was a combat pilot throughout Desert Storm in 1990. He undertook four operational tours and 490 arrested landings on an aircraft carrier. To this day, he commands huge respect among former military colleagues. During an aerospace gathering in New York in June 2006, shuttle commander Mark Kelly spotted Binnie darting out of a conference room after a presentation. Kelly, a veteran of orbital flight, had recently returned from his stint as pilot for the space shuttle Discovery STS 121, but he was still keen to catch up with an older, former colleague. 'Brian's a great guy. We flew together on the USS *Midway* during the first Gulf War. I was on A6Es, while Brian was on F18s. I've been following his exploits in SpaceShipOne and I know Brian as one of the best guys I have ever flown with,' said the NASA man from New Jersey.

When Binnie's distinguished navy career came to an end in 1998 he moved into the commercial world back in California. 'I came out to the Mojave as a test pilot for Rotary Rockets. It was a very ambitious project and we hired Scaled Composites to build the vehicle structure,' he recalls. 'So I was employed by Rotary but spent a good deal of time at Scaled working with the people and I got to know Burt Rutan.'

They became famous golfing partners, battling it out on the dusty nine-hole Mojave Desert Golf Course, which was basically an expanse of scrubland fairway and some hard-tufts of dried grass for greens. It wasn't St Andrews, but great for sand-wedge practice. And for Burt and Brian it gave them a chance to examine the science of trajectories. Secretly, Burt's ambition was to design a more aerodynamic golf ball and carbon-fibre golf clubs for the average golfer.

After the Rotary project fizzled out, Burt approached Binnie asking him to come and work at Scaled Composites as a test pilot, a move that was to lead to his career highlight – the X Prize.

Since then he has been feted and honoured around the world.

As the last pilot of SpaceShipOne, he is the best person to explain the experience for the suborbital space tourist. The top of a spaceship trajectory is known as the apogee. How you arrive there will be examined shortly, but the often under-reported down-slide, or re-entry, is worth looking at. 'And it's really quite interesting how it sneaks up on you. Re-entry from space, that is,' says Binnie.

Imagine, there you are: unencumbered and weightless, taking in a peaceful panorama stretching more than a thousand miles in each direction. You've just come through the ride of your life, a thundering ninety seconds, courtesy of a fail-safe hybrid rocket motor. As that motor switched off, you were instantly transported to a whisper-quiet realm, immersed in the surprisingly good feeling of weightlessness.

Binnie explains: 'While the ride up gave you a sense of zip even Superman would appreciate, without the motor's thrust, the ride became so smooth you couldn't detect the ship's velocity swinging from that of a speeding bullet to zero and back again.'

Your first inkling that you've still got to get down comes with a soft pinging of the cabin floor. Until then, the silence has been so profound that you literally could hear a pin drop.

'Re-entry sounds a lot like driving into a rainstorm. The first few sounds are isolated and sporadic, but they build. The drops become a noticeable sprinkle, then an insistent shower, turning into a thundering Niagara!'

But before you go over the falls, you are gently reminded that your escape from the bounds of Earth was only temporary. Some of you will elect to get back into your seats but others will dally just long enough to resign themselves to the floor. And this is just fine, since your Virgin Galactic spaceship is carpeted to provide a soft and safe haven for the tardy.

'And as your body begins to sag under the relentless pull of gravity you are grateful to be able to lie down flat and absorb relatively easily what would otherwise be a punishing ride.'

Outside, SpaceShipTwo has reached 'stagnation point', where the racing molecules of air are stopped in their tracks as they bombard the speeding spacecraft as it begins its descent. This is when the intense heat builds up. The temperature produced is somewhere approaching 1,000°C. But the profile of your flight means that SpaceShipTwo is exposed to this intensity of heat for a very short duration, so the actual heating transferred to the structure of the vehicle is low.

SpaceShipOne's feathering concept was helped because the spaceship was so light, compared with the space shuttle or the X-15. This is a ping-pong ball compared to a rifle bullet. The streamlining of these heavier vehicles mean they still go fast when the atmosphere gets thicker – this is where thermal re-entry becomes a problem. Rutan wants to cut down what is called the ballistic co-efficient – the drag of a ping-pong ball is much higher than a bullet – so as soon as it hits the atmosphere again, it starts to slow down. In feather mode, the spacecraft literally folds in half as the tail turns upwards – which increases drag.

Another of the benefits is that, just like a badminton shuttlecock, the spaceship aligns itself and assumes the right attitude to hit the atmosphere. In its feathering mode, SpaceShipOne was so stable that the pilot had nothing to do until around 60,000 ft, when he would start to glide back down to the airport.

'We measured thermal couples on SpaceShipOne and temperatures coming up to 250, and in some cases 350 degrees, and that is really a benign environment for the composite structures. So it is not going to be an issue when flying on SpaceShipTwo,' says Binnie.

The Scaled Composite designers are also working on heat protection on the outside but – because this is a suborbital flight that does not reach orbital velocity – it will not be a safety-critical feature of the vehicle as with the space shuttle.

At five, six then nearly seven Gs, you are subjected to the full force of what sounds like a tornado invading the cabin. 'What's so unusual about this "welcome home" is that, despite the noise and the Gs – especially compared to the ascent – it has been the smoothest ride of your life,' says Binnie. Until, that is, you become subsonic again around 75,000 feet.

The noise and G-forces finally disappear, leaving you with the impression you're riding in a falling bathtub. 'In the thicker bite of the atmosphere the ship's feather re-entry configuration has served its purpose and the craft complains and bucks like a frumpy bull demanding to be transformed once again into its alter ego – an aerodynamic and graceful creature from which you can enjoy the leisurely glide back down to touchdown.'

Then it is a safe and easy ride back to the spaceport.

Earlier, as a passenger on the VSS *Enterprise* you'll ride underneath the mother ship on your way to the release point and have an hour or so to reflect on the meaning of all the academic and hands-on training you've received over the previous few days. You will have been through zero-G training in a converted airliner that has provided you repeated peek-a-boo exposure to that magic environment. You will remember how easy it was to over-control even your simplest intentions. In zero-G you are Superman and it takes some time to recalibrate your muscles to their new-found prowess.

You will have been through some acrobatic and G-tolerance training, very quickly learning how humbling that can be. The spaceship will afford you better protection with its padded reclining seats, but you realise the potential is there for some wild movement and you work hard to try and acclimatise yourself to its demands.

You will have earned the equivalent of a doctorate in all the physiological nuances of the space arena and will have sat through enough rehearsals in the simulator to know when things are going to happen and what the emergency procedures are in the event of a problem.

So as you ride underneath the mother ship all of these experiences and thoughts will occupy you, along with updates and various reminders from the crew. The atmosphere is bristling and electric. You are tense, excited and nervous. And those core feelings only grow in intensity as you climb higher and higher up into the delirious blue.

You listen on the headset to the last-minute communications between the spaceship, mother ship and mission control. All systems have been checked green and go. It's ten seconds to the release. You take another sip of water because the adrenalin has kicked in and is starting to dry your mouth. You follow this with a deep breath and brace yourself.

The pilot calls '3-2-1, release'. The little 'clunk' you hear that follows is almost an anticlimax as you realise you are no longer tethered to the mother ship, but free.

Towards the front of the cabin you see a yellow light blink on, indicating the crew has armed the rocket motor. It is quickly followed by another light indicating that the fire sequencing has been activated. A large numeric display shows the number '5'. This is the number of seconds you have left before the rest of your life changes forever. This time you take a really deep breath – without thinking about it all the muscles in your body are tense and your heart is off to the races. This is all to be expected and normal, you remind yourself. The number changes to '4'; a quick look to your immediate neighbour reveals similar symptoms. Now '3'; time for a quick prayer, even if you don't pray. Next '2'; oh my God! Finally . . . '1'; you're off. If re-entry is subtle, then its polar opposite is the ascent. It's anything but.

Imagine an aircraft-carrier catapult shot that continues unabated. It accelerates you to 150 knots in about 2.5 seconds, and most would agree it is a good kick in the pants. But your magic carpet ride's 'cat-shot' will persist. At the 5-second mark, you will be at 300 knots, and at 7.5 seconds, 600 knots. Soon thereafter, supersonic! You are climbing at 150,000 ft per minute – 1,700 mph – straight up.

Brian Binnie explains that SpaceShipOne's profile can be divided into four phases.

'The first is affectionately known as the "Holy Shit!" phase, which kicks in as that rocket motor wakes up and literally sweeps you away.'

There are instant G-forces pushing through your chest. There is huge noise, vibration, shaking and surging. Your senses will be pegged and despite all the training and the lectures you've sat through you will have little experience to reassure you everything you feel coursing through your veins is 'normal'.

'With every nerve of your being focused on the flood to your senses you will know without a doubt you are very much alive,' continues Binnie. 'The dominant thought if you could focus on it would be screaming, "Holy Moly, I'm riding a rocket!" '

The second phase occurs during the approach to Mach 1 after about ten seconds. Shock waves are now dancing on different parts of the vehicle at random times, causing rather violent and dynamic reactions. If you can pull your head up enough to look out of the window it will show you to be approaching about 60 degrees nose up, although it feels to you like you're pointed straight up. But whatever the angle, there is no doubt you are going up – and rapidly.

The third phase lasts about 45 seconds, while the vehicle accelerates from Mach 1 to Mach 3. The ride smoothes out dramatically and the suddenness of the Gs, noise and vibrations have fallen to the background; you will be surprised at how quickly you've acclimatised to this new environment. Phase three ends in a rather dramatic fashion. Like almost everything about this ride, just about the time you think you've caught up with it, it reveals to you another aspect of its nature.

The rocket motor now demonstrates the combustion properties of the liquid-to-gas transition. This is when the irregular oxidiser flow into the combustion chamber causes about five seconds of significant airframe shuddering. It is a rude wake-up call as it announces that the final phase – and the hardest part for the pilot – is about to begin.

For the pilot, the phase-four end game requires trying to 'feel' a balance between limited aerodynamic control and the capricious tendency of a now-dying rocket motor to upset all the hard work accomplished to this point. The g-level has diminished significantly and the rocket motor is now running on the remnants of its fuel.

Binnie describes this as sounding 'as though you are pulling on the tail of a possessed cat. This ungodly screeching announces the final struggle between the decaying rocket and the pilot's ability to balance its unpredictable performance in the wispy upper atmosphere.'

And finally here comes the magic in your carpet ride.

'When that shrieking rocket motor is finally shut off, you literally step across a threshold into another realm, where beauty and blessed peace and quiet reign, graced by the instant karma of weightlessness.'

And, my God, that view! The black, foreboding void that is space is magically revealed as though someone has pulled back a stage curtain for your eyes only. This vast presence, looming and yawning through the windows, offers both menace and mystery.

Below is a reassuring comfort – a 1,000-mile horizon that reveals a magnificent splendour of mountain ranges, coast lines and weather patterns normally only seen on the evening news. And separating space from Earth is an improbably thin, bright, electric-blue ribbon that is the atmosphere.

Everywhere you look is: 'Wow!'

And everything you feel is: 'Wow!'

And as you drift over to another window for a different view, you realise with childlike wonder that you are in a spaceship in space!

'The VSS *Enterprise* has just about wrung you out emotionally in a full-spectrum experience that will likely become a centrepiece of your psyche and dominate your perspective from that moment on. I believe you will feel enriched, enlightened and – most important of all – happy,' concludes Binnie.

But with four more long minutes of magic still to be absorbed you glide lazily over to a companion to share and soak up the euphoric reward of all your hard work and relish the joy it brings!

Binnie's description is a brilliant synopsis of the suborbital flight. He has been through this – and he is determined that many more people should experience nature's shock and awe. It's certainly something people are willing to pay for.

12. A SPACEMAN JOINS THE SHOW

Winning the X Prize with a tiny experimental spacecraft was a significant turning point – but now a different kind of hard work would begin inside Scaled Composites. A whole new way of working with procedures and exacting systems would be required to build a robust and commercial spaceship capable of carrying six paying passengers safely into space and bringing them home again. The science and the innovative technology were proven – now they needed to be scaled up to meet the exacting regulations of the US Federal Aviation Authorities and the insurance companies.

Virgin Galactic now needed to find a heavy-lifter with a space pedigree who could talk Burt's language, yet look after their interests and considerable investment. This needed a cool diplomat, a mediator and an egghead with a brain that could trade trajectory equations at the blackboard with Burt Rutan. Not an easy CV for the headhunters to fill. But George Whittinghill, the chief technical officer for Virgin Galactic, fitted the bill – and with some to spare. He is responsible for the SpaceShipTwo project and he has been working alongside Scaled Composites with the Spaceship Company.

The entrepreneurial culture at Scaled is a reason for its continued success. Sleeves are rolled up to get the work done and there is a small and effective chain of decision-making. 'One of the strengths of Scaled is that we are a very flat organisation. There is no one in the structure who can't apply their expertise quickly to a problem, even on the business side,' says Rutan.

The management theory textbooks would call Burt's team 'fast prototypers'. Fast prototypers are an interchangeable group of engineers and technicians who can rapidly define a technical or mathematical problem. In Scaled's case, once they understand the issues – and how they relate to other parts of the plane – they can reshape the vehicle very quickly to get things moving again – from the concept, to the computer, to the design of the tooling, then next door to build it in the workshop, then back to the computer again to test and redefine. This is a novel way of creating airplanes and aircraft systems. But Scaled have used this practical philosophy to build over thirty complete aircraft – all with different characteristics.

'Scaled Composites can outbuild any similar aerospace company in the United States – even the big boys. This isn't arrogance, it's a fact for us,' says George Whittinghill. 'We are building a commercial spaceship from the outside in, rather than inside out. This is the Scaled way of working.'

George admits he was a space geek from a young age. 'When I was kid we were in the prime of the space race and NASA had what it appeared, from my perspective back then, no intention of backing off. We were going to the Moon regularly and there were plans to go well beyond that. The public interest was swelling and space was a common topic sitting around our dinner table, and there were movies and TV shows to watch too. So it was a very exciting time when I was an impressionable boy growing up in California. We all thought, boy, aerospace is just going to blossom.'

This attitude attracted many of the brightest science-struck kids in the US of the 1960s and 1970s – and it let them down big time. After enticing so many of the nation's top brains, the outlets for this imaginative and technologically advanced cohort dried up like a Mojave river bed. For Whittinghill, the one oasis was the space shuttle.

'The shuttle was pretty good. As a system it is complex, but very capable for both manned and cargo payload, but unfortunately it has had a couple of accidents.' Indeed the *Challenger* accident in January 1986 destroyed a lot more than a huge craft with seven astronauts on board. It obliterated the dreams of many.

'Of course, Mike Griffin has changed the whole architecture of the way that NASA proceeds. It is definitely back to the Moon again. The shuttle missions have been focusing on space operations and helping to build the International Space Station with the Russians. They have been building an understanding of what it is like to really live, work and assemble things in lower orbit. I think the legacy of the shuttle is to understand how to integrate systems in orbit. There was a lot of space walking too – it has been tremendous for that.'

So while NASA and the exploratory space missions will continue, the new generation of space-tourist companies, such as the Spaceship Company, will concentrate on building less elaborate and simpler crafts to take paying passengers into suborbit. This has become Whittinghill's speciality.

An expert in manned spacecraft systems as well as hybrid rocket propulsion, he is a softly spoken scientist. 'I was working as a consultant and was introduced to Alex Tai by a friend of mine, the astronaut Mike Foale.'

Foale, born in England but now an American, is a record-breaking astronaut: in six missions into orbit he has logged over a year of his life – 374 days, 11 hours and 19 minutes. He was selected as an astronaut by NASA in June 1987 but prepared for flights on the Russian space station *Mir*, training at the Cosmonaut Training Centre in Star City. He served on STS-45, STS-56, STS-63 and STS-103, was a flight engineer on Mir 23 and Mir 24 (going up in STS-84 and coming home on STS-86), and served as the International Space Station's commander. Foale holds the American record for time spent in space.

Top right: Yuri Gagarin: the first space tourist. The Soviet cosmonaut became the first human to blast off into space in April 1961. 'I see the Earth. The loads are increasing. Feeling fine,' he said after lift-off. His orbit of the Earth in Vostok-1 forced America to enter a manned space race. (Picture: NASA)

Top left: Rocket scientist Wernher Von Braun is briefed about satellite orbits by Dr Charles Lundquist at Army Ballistic Missile Agency, Huntsville, Alabama, in 1958. (Picture: NASA Marshall Flight Center)

Above right: A forerunner to SpaceShipOne: the Bell X-1E airplane is loaded under its mothership, a Boeing B-29, in 1955. The X-planes were launched from beneath the plane and fired towards space. (Picture: NASA)

Right: Neil Armstrong, the first man on the Moon, pictured next to the X-15 after a research flight in 1960. The X-15 was a rocket-powered aircraft, flown from June 1959 to October 1968. It set the world's unofficial speed and altitude records. The X-15s made 199 flights. (Picture: NASA)

Above: The Mercury Seven astronauts: the first American astronauts announced in April 1959, six months after the creation of NASA. Front row (left to right) Walter Schirra, Donald Slayton, John Glenn, and Scott Carpenter; back row, Alan Shepard, Gus Grissom, and Gordon Cooper. (Picture: NASA)

Right: President John F Kennedy makes his inspirational address before 35,000 people at Rice University in Texas in August 1962. 'We choose to go to the Moon in this decade and do the other things, not because they are easy, but because they are hard,' he said. (Picture: NASA)

Above: The X-15 under the wing of a B-52. It was air-launched from the B-52 at 45,000 ft and at a speed of about 500 mph. This photo was taken from one of the observation windows shortly before dropping the X-15. (Picture: NASA)

Right: Ground-based rockets became the norm for space flight. Here Apollo 11's Saturn V rocket lifts off from Kennedy Space Center with astronauts Neil Armstrong, Michael Collins and Edwin Aldrin on board on 16 July 1969. (Picture: NASA)

Above: Edwin 'Buzz' Aldrin inside the Apollo 11 Lunar Module during the Moon landing mission in 1969. Picture taken by Neil Armstrong. (Picture: NASA)

Below: A view of the Space Shuttle Atlantis leaving the Mir Russian Space Station in 1995. This image was taken during the STS-71 mission by cosmonauts aboard their Soyuz TM transport vehicle. (Picture: NASA)

Above: NASA successfully completed its first rendezvous mission in space with two Gemini spacecraft in December 1965. This photograph, taken by Gemini VII crewmembers Frank Lovell and Frank Borman, shows Gemini VI, crewed by Walter Schirra and Thomas Stafford, in orbit 160 miles (257 km) above Earth. (Picture: NASA)

Left: The super-rich space tourist: Dennis Tito, the first private citizen to visit the International Space Station, shares his experiences at the 40th Space Congress at Cape Canaveral, Florida, in May 2003. (Picture: NASA)

Below: The record-breaking, nine-day Voyager flight in December 1986, which flew 26,678 miles non-stop around the world without refuelling. The Voyager aircraft circles before landing at Edwards Air Force Base, California. Dick Rutan and Jeana Yeager piloted the aircraft, designed by Burt Rutan. (Picture: NASA Dryden Center)

Left: The experimental Proteus aircraft, designed by Burt Rutan, served as a test bed for NASA flight tests. The tests were flown over southern New Mexico in March 2002. (Picture: NASA Dryden Center)

Left: The new space pioneers: aircraft designer Burt Rutan, X Prize founder Peter Diamandis and Paul Allen, the billionaire co-founder of Microsoft who funded SpaceShipOne. (Picture: X Prize Foundation)

Left: Burt Rutan shows the unique 'feathering' re-entry system for SpaceShipOne. (Picture: Virgin Galactic)

Below: SpaceShipOne tucked underneath its mothership, WhiteKnightOne. Virgin Galactic's spacecrafts, taking six passengers into suborbital space, will be a larger version of both of these craft. (Picture: Virgin Galactic)

Above left: Actress Victoria Principal, the former *Dallas* star, talks about her desire to go into space. (Picture: Virgin Galactic)

Top right: Governor Bill Richardson and Sir Richard Branson sign a deal for Virgin Galactic to become the first major tenant at Spaceport America in New Mexico. (Picture: New Mexico Economic Development Agency)

Above left: The founding astronauts gather for a photo call with WhiteKnightOne. (Picture: Virgin Galactic)

Left: The launch of SpaceShipTwo's interior design and passenger cabin, below SpaceShipOne, in New York in September 2006. Note the difference in size between the two spacecraft. (Picture: Virgin Galactic)

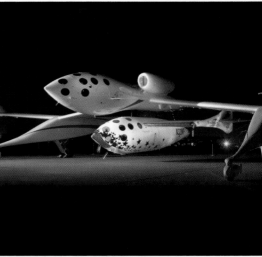

Top right: The future of space tourism: a mock-up of Virgin Galactic's spacecraft. (Picture: Virgin Galactic)

Top left: Virgin Galactic founding astronaut Lina Borozdina who has re-mortgaged her Californian home to buy a $200,000 ticket into space. (Picture: Virgin Galactic)

Right: Virgin Galactic President Will Whitehorn who has driven the project in the United States. (Picture: Virgin Galactic)

Bottom right: A spaceport beckons: this site in the New Mexico desert will be converted into a spaceport by 2010. (Picture: Virgin Galactic)

Below: Mars – the future. (Picture: NASA)

While still an active astronaut, he has been NASA's explorations administrator with a keen interest in the idea of suborbital space. In July 2005, Foale and NASA colleague Kenneth Cockrell visited Rocketplane in Oklahoma to discuss space elevators and personal air-vehicle technology. They had a constructive session with Bob Seto, Rocketplane's chief engineer. Foale even raised the subject of NASA hiring Rocketplane's XP suborbital vehicle.

'Dr Foale was interested in using the XP for astronaut training, procedural training and non-Earth gravity experience. We could provide three to four minutes of weightlessness and much more time at Moon and Mars gravities,' Seto told *Flight International*.

Foale has also been taking an interest in the work in Mojave. So there was now increasing commercial interest in what was happening in the suborbital sphere. Mike Foale and George Whittinghill were old pals from their time at NASA's Johnson Space Center in Houston and had kept in touch over the years. Whittinghill was signed up to become the liaison officer between Virgin Galactic and Scaled. He sees his job as 'being able to keep the Virgin team appraised of the technology and what is going on'. He has a practical pedigree that builds confidence.

An aerospace engineer with a Bachelor degree from MIT in Boston in 1978, George's first job out of university was working for McDonnell Douglas in ballistic missile defence in the South Pacific. He was newly married and working on the Kwajalein Atoll, in the centre of a triangle from Japan, New Zealand and Hawaii. It is north of the Equator and in the middle of nowhere. 'It was a tremendous amount of fun but then I went back to graduate school in Boston. I don't know why I went from the beautiful tropics back to snowy Boston, but I did!' he recalls.

Armed with a Masters in space propulsion, he was able to join NASA at Johnson. 'I went to work as a civil servant for a few years and I was training astronauts and flight controllers. I then worked on the payload side. But after the *Challenger* disaster I decided to go and work in commercial space with a small company called Space Industries in Houston.'

The company was set up by former astronauts and physicists from the original Apollo era, such as Joe Allen. Joe missed out on actually flying on the Apollo missions, but flew twice on the space shuttle, including its first fully operational flight. His foil was the diminutive Max Faget, a NASA stalwart who oversaw the design and development of the original Mercury, Gemini, Apollo and space shuttle vehicles. Faget died on 9 October 2004 aged 83, a legend in the industry. Also on the team was Caldwell Johnston who worked on the Mercury capsule. These guys knew what they were talking about. The company wanted to build a mini space station – a 'man-tended free flier' – called the Industrial Space Facility. This would process high-value pharmaceutical and electronics in zero gravity, but they needed the shuttle to service it.

Whittinghill worked at Space Industries for a number of years and even encouraged Cliff Kurtzman, the chief executive officer of ADASTRO Inc., a leading rocket science company, and also an MIT space engineer, to join him, but it didn't survive. The science and engineering was sound, but after a major acquisition the funding became a problem. Kurtzman said later it was a company with two many high-powered chiefs and not enough Indians. The company was waiting for NASA to issue a proposal for its service but it turned out to be a bid that was too close to what the International Space Station was doing in terms of science research. NASA decided not to go for the Industrial Space Facility, which even George conceded was a good call because the ISS is so much bigger and more capable. Gradually, the company moved away from its focus on space and went public in a flotation, changing its name to Veridian.

'I decided after that if you are really going to commercialise space you have to start from the ground. If you can't get the stuff up there cheaply you are not going to make a business case up there. So I looked around for companies that were in commercial space transport and at the time there were two:

Orbital Space Sciences Corporation and the American Rocket Company,' George says.

OSSC consisted of only three people at the time and ARC was using a brand-new system called 'hybrid rocket' technology. The company was based in California and built everything from top to bottom. 'I decided that these guys controlled everything and they had the greatest chance of gaining access to space, so I went to work for them.'

George joined the American Rocket Company in 1989, where he became director of special projects. The company managed to put a rocket called Set One on the launch pad at the Vandenberg Air Force Base in California, home of the Air Force Space Command. It wasn't impressive though. It failed because of a main valve malfunction, much to the embarrassment of all the ARC people. But there was one positive outcome. It proved to the world that hybrid rockets were safe. 'Not the way we wanted to,' concedes George, 'but it didn't blow up. That was always our claim to fame.'

The company struggled along for another six years, building the world's biggest hybrid motor and firing it: a huge beast of a 32 ft long, 6 ft diameter motor with 250,000 lb of thrust. But in 1995 the business was under pressure and went bankrupt. And here another competitor to Virgin Galactic entered the fray – and showed what an incestuous place the commercial space travel industry was becoming.

Jim Benson is the founder and the largest shareholder of SpaceDev, which he started as a space-technology development company in the San Diego suburb of Poway back in 1979. In 1998, SpaceDev acquired all the intellectual property that had been generated by the American Rocket Company. Over the previous eight years the company had performed over 300 rocket motor tests, from 15 lb to 250,000 lb of thrust, and was the leading hybrid-rocket company in the world. Benson now had a head start and enhanced and improved the designs. So when Burt Rutan was looking for a motor, he knew where to go. And SpaceDev began working on the hybrid-rocket motor technology for SpaceShipOne in

January 2000. SpaceDev supplied all the rocket motors for SpaceShipOne's X Prize flights.

Benson is a former geologist who formed CompuSearch and then ImageFast, building them up, selling them and retiring at fifty. He then spent the next eighteen months looking at what to do next.

'The space industry needed revolutionising. It was just moribund and bogged down with the old mainframe mentality dominated by the large aerospace companies. I really became the first hi-tech entrepreneur to move into the commercial space industry. I've had a lot of followers since then.' Jim is pretty cocky about his achievements. 'There have been a lot following in my footsteps, but so far SpaceDev has been the only successful company.'

George Whittinghill knows Jim, and he too enjoyed the entrepreneurial aspect of the space business, working with several start-up companies as a consultant. 'I was always trying to figure out ways of getting low-cost access into orbit. It has to start there.'

Eventually he landed with the Space Launch Corporation, with founders Michel Kamel and Jacob Lopata, which won some funding for a series of projects called Rascal. 'That's when I really started working with Burt Rutan. It consisted of a high-altitude supersonic airplane that would zoom and coast to about 180,000 feet, deploy a rocket with a very large payload and the rocket would go into orbit.'

This had the potential of being the lowest-cost method of putting a payload in space. The Rascal was a two-stage rocket system designed to propel the craft into orbit, a high-altitude airplane launch similar to the X-15.

'So we had Scaled on contract to build the airplane, called the MIPC-powered Vehicle, which was an acronym for Mass Injection Precompression Cooling. This is propulsion technology to enhance the turbo fans to operate at much higher speed and altitudes than they normally do.'

In the upper atmosphere, the air thins to nothing before it hits space, and engines need oxygen to combust. While the air is thinner, the engine gets much hotter as the craft goes much faster. The MIPC system was able to cool the air down and then force more of it through the engine. This allowed a triple-fan engine to operate at extremely high altitudes – around 100,000 ft – where there was virtually no air. The aircraft would go straight up to its peak and, close to the zenith of its ballistic arc, a door would click open, the payload would be ejected and this would go all the way into orbit.

Scaled Composites were responsible for building this vehicle and George worked closely with Burt Rutan, as chief technical officer for the Space Launch Company, based in Irvine, California. That lasted through phase one and two of the programmes, sponsored by the US's Defense Advanced Research Projects Agency (DARPA), but when it came to detailed design and build, DARPA decided to back out. Instead the funding went into Elon Musk's Falcon, who were also looking at low-cost launches but using a different method – pushing a two-stage rocket, powered by liquid oxygen and kerosene, up into orbit out of the back of a massive C17 transport plane.

'We were high and fast and they were low and slow,' says George. 'We were driving the cost of putting payload in orbit, with very, very low values. But our development costs were high because we had to develop all this new architecture, whereas the Falcon approach uses existing air transport in the C17, which is a lot cheaper. But what is important here is getting payload in orbit cheaper.'

Falcon won the contract and George Whittinghill went off to join Virgin Galactic. Falcon's development moved on rapidly to a different level and a launch was set for the first flight of SpaceX's 570 kg (1,245 lb) rocket on 25 November 2005, planned for the US Army's Reagan test range on Omelek Island, one of the Marshall Islands in the Kwajalein Atoll. The launch was delayed, though, when they hit a rather basic problem – the liquid-oxygen plant in the tropics wasn't able to produce quite enough cooling gas to fill the spacecraft.

By virtue of his deep knowledge of hybrid motors, it became obvious to Burt Rutan that George, by then acting as a consultant, would be a great asset. He was able to help with the SpaceShipOne project, offering some specialist input. 'It rapidly became apparent that I could help. Burt wanted to bring the hybrid-rocket motor technology in-house into Scaled Composites. He didn't want to contract it out to anybody. He wanted to be completely vertically integrated, so I agreed to help him do that.'

The bonus for Burt Rutan was that Whittinghill could also provide some valuable insight from his years working with NASA and the space community and in the systems to be used on SpaceShipTwo and WhiteKnight. 'I could also make sure that Virgin Galactic was getting what it needed too, so that it could eventually go commercial. I became the liaison and the chief technical officer.'

Imagine the scene: at Mojave there is a space-team meeting to discuss a significant development. Alex Tai and George have arrived with Will Whitehorn, John Peachey and Burt in the chair. Sitting alongside are Doug Shane and Brian Binnie. The meeting is informal but there is a sense that something significant is happening.

'I'd like to introduce you to something I've been working on,' says Burt.

He opens up a black ring-backed folder to unveil a large computerised drawing of WhiteKnightTwo in various cross-sections. The main graphic shows a sleek, futuristic jet plane with two hulls, strikingly similar to the format used for Virgin Atlantic's GlobalFlyer. But this design is for a much larger four-engined plane with two jets at the end of each hull.

'As you all know, we've been trying to solve this issue of how we release SS2 from underneath the mother ship as cleanly and safely as possible. We've run all the tests on this and we believe this twin-hulled approach is a significant shift,' he says.

'This is not going to look like the spaceship in your animation, Will,' Burt warns.

'I can see that, Burt. But it looks incredible,' remarks an astonished Whitehorn.

For Alex Tai and George, it is a sensible solution to a key issue of scaling up to a fully operational commercial spaceship.

'Reliability and safety are the prime factors for the whole project. I want to make sure that at all regimes in flight we are covered in terms of malfunctions, failure recovery and that the operations are safe and the systems are designed to be robust and failure tolerant,' chips in Whittinghill.

'Why sure,' replies Burt.

WhiteKnightTwo, designed by Burt Rutan, is a brand-new plane and its purpose is to support and complement Space-ShipTwo. It has very long, thin wings to provide the necessary lift at extremely high altitudes. It is a stylish airplane – with typical Rutan trademarks.

Alex Tai is keen to understand what the space pilot's experience will be like. Already he has a briefcase full of CVs from Virgin Atlantic and other airline colleagues wanting to be considered to fly – and there are literally dozens of other experienced ex-service pilots who have heard about the specification and the excitement and want to get in early.

'For the pilots it will be like driving a sports car. It won't be slow and plodding like a Boeing 747 jumbo jet. The acceleration of the rocket ship under power is about 3 Gs. It is a very nimble and fast-moving ship that will be fun for the pilots,' explains Burt.

'But the pilots will really have to fly these planes. It's not simply push one or two buttons and it's on autopilot. The pilots will be on the stick, looking at their instruments and making all the instant decisions,' he adds.

But when would test flights begin? asks one of the team. Whitehorn is uncertain. There is still the serious issue of whether the US government will allow sensitive technology to be exported to Britain. This still needs a proper resolution.

'I really want Virgin Galactic to be in service in either 2008 or 2009. We will launch as soon as our safety assessments and training dictate we do so, and not a day before. But our launch date estimate assumes prompt clarification of the US Government technology licensing issue,' he says.

The longer it remains unresolved, says Whitehorn, the more it could adversely impact the projected launch date. And, as far as making money is concerned, Virgin Galactic's business plan projected profitability within the fourth or fifth year of operation.

'Importantly, this estimate assumes five spaceships, two launch aircraft or mother ships, and two launch bases in the United States. If the schedule for deploying any of these assets slips, it would negatively impact our target date for profitability,' says the Virgin Galactic president.

There are many more important issues still to be thrashed out. But, at last, here was a system that was going to fly safely. For Burt and Will it was now a case of winning over the regulators.

13. HYBRID ROCKETS ARE COOL

Professional astronauts can be remarkably blasé about life and death. It's an occupational hazard. They call sitting on top of a rocket filled with hundreds of tonnes of explosive fuel 'riding the stack'.

A rocket launch is the most powerful force a human being can ever experience – and survive. It is an explosion of titanic power, with kerosene, hydrogen and oxygen igniting in a ferocious fireball. The Saturn V rocket produced an enormous 7.6 million pounds of thrust.

It is a memory that will last forever for the space tourist. But it does get very bumpy – and noisy.

Neil Armstrong reported that his flight to the Moon was a 'magnificent ride' on the Saturn V rocket, but the truth was he would have preferred something a bit gentler. 'In the first stage, the Saturn V noise was enormous, particularly when we were at low altitude, because we got the noise from seven and half million pounds of thrust plus the echo of that noise off the ground that reinforced that.'

He said the first thirty seconds were very difficult to hear anything over the radio – even inside the helmet with the earphones on.

On the space shuttle, the astronauts lie on their backs with their knees up. Four people are upstairs with three more in the mid-deck. All they have to do is sit back and enjoy the ride. With ten seconds to go they hear the noise and rumble of the water deluge system emptying into the launch pad to absorb the heat and shock. As the countdown goes on, the main engines are throttled up before the solid rocket booster is ignited. It kicks in at 2 Gs with the whole cabin rattling, like it's suffering an earthquake. Then seven million pounds of thrust drive the craft from zero to 17,500 mph in less than eight minutes. Jeremy Clarkson, eat your heart out.

The rocket engine of the VSS *Enterprise* is one of the safest ever made.

At Scaled Composites, there is a propulsion team of six engineers plus technicians, who are working on the main systems of firing paying passengers into space. A project engineer is responsible for each of the following areas: SpaceShipTwo, WhiteKnight and rocket-motor developments. The overall programme manager is Jon Karkow, who oversees the three project areas, while the chief aerodynamicist is Jim Tighe. The project engineers have the authority to bring a programme forward and they are in constant discussion with Burt. If they hit a challenge, they go and see Burt Rutan and he provides advice.

Simplicity is the key. The fewer parts – the less chance of any failure. This is the Rutan doctrine, through and through.

The propulsion team – responsible for the rocket – have been using advanced computer modelling to work out flight dynamics, or 'drag', to assess the best range and design for SpaceShipTwo. And comprehensive testing at Scaled Composites is now under way. Not a single paying passenger will fly until there have been nearly a hundred test flights to ensure that every component meets the rigorous demands of the US Federal Aviation Administration.

George Whittinghill admits there is a tremendous amount of calculations and maths involved in the work. Inside Burt's office there is a whiteboard and felt pens for these theoretical sessions where the teams have been thrashing out issues. Burt isn't into long, drawn-out meetings and processes. 'It very quickly gets down to the nitty-gritty. Things get done quickly,' said George.

The key to getting into space is the hybrid rocket motor – and it is inherently safe. So how does it work?

Fundamentally, all rocket engines have an oxidiser and a fuel. The oxidiser creates the oxygen for when the rocket has escaped from the Earth's atmosphere, where there is no air. There are three basic kinds of chemical engine types, so let's look at each in turn.

First, the solid propellant. These are like tactical-grade missiles where the oxidiser is mixed in with the fuel all in one motor case. This is lit and it zooms off up into the air and cannot be extinguished – it burns all the way to the end. Large rockets, such as the space shuttle and the Arianne 5, use solid strapped-on rockets like this. It's a highly reliable way of boosting payload into space, but it is very messy to the environment. And if it starts to go wrong it is almost impossible to stop it. It can blow up with tragic consequences.

The second type of chemical rocket propulsion uses liquids. Here an oxidiser tank and a fuel tank are used, typically containing liquid oxygen and kerosene – used for commercial airplanes. The fluids are either pushed or pumped into a combustion chamber and injected into a mix and then ignited. The advantage of this technology is that a much wider variety of propellants can be selected. Space engineers can get much better efficiencies than with solid-fuel rockets. Liquid hydrogen and liquid oxygen technology is used by the space shuttle's main engines. But this is a highly complex system with cryogenic propellants, which are extremely cold, being pumped into a chemical cycle where the gaseous temperature and ferocity of heat is greater than the inside of a blast furnace. There is a huge requirement for extreme heat and then cold, with metals and alloys robust and flexible enough to take the extremes of expansion and contraction. This makes it all very expensive to develop.

Finally, and somewhere in the middle, is the hybrid system, favoured by Burt Rutan and Virgin Galactic. This is still a dual-propellant system with a liquid oxidiser but there is also solid fuel. The solid fuel lines the case of the rocket with a nozzle but the fuel has no oxidiser in it. Instead the liquid is injected in the head end of the motor and then ignited. The surface of the solid fuel turns to gas, reacting and combusting inside.

'The beauty is that because the propellants are separated physically and by phase – meaning one is a liquid and one is a solid – they cannot intimately mix in the event of a leak or something going wrong. So consequently, they cannot explode. They can't detonate and they are very failure-tolerant. If the fuel inside a motor case cracks, it is not catastrophic the way it is with a solid rocket propellant,' says George Whittinghill.

The other aspect for the fledgling space-tourism industry is that hybrid rockets are inexpensive to make. Once all the engineering and design has been done, being able to crank them out on a production line is relatively simple. Rubber is used as a fuel. The latex is poured into a mould and a catalyst and curative are added. Chopped-up tyres could be used but they do have additives, such as sulphur, which would make it less effective for hybrid rockets and would produce sulphur dioxide as a by-product. So normally clean chips of synthetic rubber are being used, a solid-fuel log of hydroxy-terminated polybutabiene (HTTB). It then just sets like a thick black log, twenty inches in diameter, which is inserted inside the motor. The log has four-inch thick walls and a four-inch partition that runs the length of the log creating four pie-shaped cavities. This log can then be encased in steel or a composite case.

Once the igniter motor starts the rubber burning, the nitrous oxide is added under pressure, producing a serious flame. The expanding gas accelerates through the nozzle and provides instant thrust. Whoosh!!

'Composite is the favoured way to go because it is very lightweight and very tough. And because the solid fuel lines the case on the inside, it acts as an insulation. You have 5,000-degree temperatures on the inside but there is all the fuel between the intense heat and the outer motor casing. The fuel slowly burns away, so the side of the case never feels the fierce combustion temperatures. The case does not see extreme temperatures until the very last moment when you are done,' Whittinghill explains.

Scaled Composites has put a thin layer of extra insulation, so the case doesn't deal with any extreme temperatures. 'It's beautiful and very nice,' says Whittinghill. 'So now you have a hybrid which is extremely safe at very low cost and with an efficiency that is in between solids and liquids. So it is ideal technology for us.'

The rocket motor is wrapped in electrical wire, yet another safety feature. If a hole appears in the rocket case – which is highly unlikely anyway – it will burn a wire in half and that in turn will shut off the valve that switches off the rocket. A fail-safe lifesaver.

SpaceShipOne had a synthetic rubber-based fuel but Space-ShipTwo is likely to be less costly and easier to produce. There is a whole panoply of fuels to look at and Scaled has tested quite a few and are in the final process of selecting the best one for commercial use. The rocket motor is designed to give just enough push – technically known as Delta V or velocity – to tip the craft into suborbit. After this the motor shuts down and the spaceship coasts into space for a few minutes. There is not enough power to take the craft into orbit and stay there. It reaches the top of its arc and then starts to fall back down again. Just like tossing your keys in the air, once they reach the top, they start to come down.

To go into orbit requires a lot more energy. Whittinghill and Rutan see this as being another major technological challenge to crack. 'It's trying to push the same amount of payload and classically that takes three to four stages. Each one, as you go down, is bigger by a factor of four. The orbital jump is still a way off for us. It's in our sights and we know we want to get there but we want to make sure that we do SpaceShipTwo correctly. This has to work safely and effectively and be a commercial success before we can credibly start looking at orbital. I think everybody is looking at orbital flight romantically and saying: "Gee, four or five minutes of weightlessness is nothing compared to ninety minutes in orbit around the world." But this is the first stepping stone.'

So what do others make of hybrid motors? Sven Grahn, a distinguished rocket scientist with four decades of experience with the Swedish Space Corporation, believes the motor is the safest way to take paying passengers into space. He is willing to give it a ringing endorsement. 'I have been in this space business for forty years and the only passenger spacecraft that I would travel in is the Virgin Galactic one. It is inherently safe – because all the others are as dangerous as hell.

'Conventional rockets are quite awesome,' says Sven. 'And really scary. We launched a telecommunications satellite in China. It was a very nice operation and it worked perfectly and I had to sit in the underground bunker for the launch. But to get there I had to walk across the launch pad. Here was this Long March 2C rocket standing with 200,000 tonnes of nitrogen tetrocide and hydrazene and I had to tiptoe past.'

Bearded Sven raises his eyebrows. 'We would not venture up in the service tower when this rocket was filled with propellant, so we pulled out all the cables from our satellite sitting on top of the rocket two days before launch. I asked: "Isn't this dangerous?" Our Chinese hosts replied: "Only if you drink it."'

But his point is well made. Rockets do blow up. Occasionally, but spectacularly.

'So when we look at Virgin Galactic's hybrid rockets we see something that is very benign indeed. I would say that it is the only way to go into space.'

He calls it a mixture of laughing gas and chopped-up garden hose. 'In a solid rocket motor you mix the oxidiser with the fuel. The fuel is usually rubber and the oxidiser is a salt – ammonium perchlorate – and this burns inside the propellant. In a hybrid rocket, the propellant is vaporised and burns outside the remaining propellant grade as it meets the decomposing nitrous oxide, which contains the oxygen. So there is a lot less heat transferred to the walls of the container of the rocket.

'If you switch off the oxidiser, it stops. You can restart it in principle but this is still a difficult thing to do at high altitude. It's not an easy combination to ignite – which is nice. Because even if it crashes, nothing happens, it is completely benign. So you can switch it off easily. Of course, you can switch off a solid rocket but it is much more dramatic. You have to let the pressure drop and puncture the motor. But, with the hybrid, you simply halt the flow of the nitrous oxide, and it stops,' says the Swede.

But the added bonus for hybrid engines is that a pilot steering a spaceship can throttle the engine by controlling the gas flow. 'It's definitely the best solution for people who want to get back in one piece,' he says.

Jim Benson, whose company built the rocket motors for SpaceShipOne, is committed to the cause too. 'Liquid and solid motors are explosive. They are both rated in tonnes of TNT equivalent. Then you have hybrid – half solid and half liquid – and you have no TNT equivalent at all. So you have the safest rocket motors in the world.'

And they are disposable, just like a Bic lighter. In Space-ShipOne, the casing and nozzle were thrown away after each flight, while everything forward, such as the plumbing, the bulkhead and the combustion chamber, remained intact and reusable.

But the Benson Space Company wants to compete full-on with Virgin Galactic – in what he dubs the Benson v Branson contest. 'Of all the companies talking about suborbital tourism there are only two that are serious. It's a David, that's us, and Goliath, Sir Richard and his guys. It's the tortoise versus the hare. This is a real race and it's going to be fun. The money is there for suborbital and then orbital, but we've got to get beyond the one rich person flying to the International Space Station for twenty million dollars.'

He is using larger hybrid rocket motors with HTTB

synthetic rubber and nitrous oxide, N_2O. And he's bullish about his own latest versions.

'Why should anyone want to buy a SpaceShipTwo? They have no evolutionary path to orbit. The Benson Space Company is choosing a proven NASA design called the HL20 Spacetaxi. It is a horizontal-landing space taxi based on the Soviet BOR4, the Boran 4. It was successfully sent into orbit four times and re-entered four times successfully. So it has a proven flight heritage, so my company is going to contract with SpaceDev, which is publicly traded. SpaceDev will be the prime contractor for fabricating the HL20 from existing plans and integrating our existing hybrid rocket motors and testing. Then we're there. We have a very simple, practical, fast, low-cost approach. That's why we're going to be first to market.'

By the end of 2008 – or early 2009 – Benson believes he will be taking paying passengers into space. 'Well, we don't have anything to do. We have the existing design and the rocket motors to test.' He is almost disparaging about Virgin Galactic, but you sense his huge respect for the power of the brand. 'We don't have to come up with a very complete twin engine that will circle for two hours to sixty thousand feet and then drop a very heavy spaceship. We just launch vertically, straight up. It will be much more exciting. You strap yourself in, you count down, then straight up and fifteen minutes later you glide back down to a runway for a safe landing where you'll see your family and friends.'

Benson is looking at five states: Florida, New Mexico, California, Nevada and Virginia.

The views of Jim Benson and Sven require a little unbiased academic input. I wanted further corroboration, so I spoke to a leading American space-industry academic, who echoed their views about the hybrids. Professor Ann Karagozian, head of the Combustion Research Lab at University of California Los Angeles, believes the rocket technology for SpaceShipTwo is highly stable.

As a former member of NASA's Aerospace Technology Advisory Committee and a member of various high-level strategic committees, including the National Science Foundation, the prospect of inexpensive commercial space flight appeals to her. Professor Karagozian describes Rutan's achievements as 'monumental', even comparing his work alongside the Wright Brothers and their invention of powered flight. High praise indeed.

'It is true that both the Wright Brothers and Scaled Composites used known technologies and implemented them in novel and creative ways into a new transportation paradigm,' she confirmed to me. 'I believe that his company's achievement is extraordinary from many perspectives, especially given that they entered the space transportation arena only two years before achieving the successful flights of SpaceShipOne. I believe that hybrid rockets such as that used by SpaceShipOne have a lot of positive features: they are reliable, controllable, have decent performance, and have better safety features than, say, solid-rocket motors.'

She explained that engineering feats usually build on prior accomplishments. 'I would not say that SpaceShipOne "evolved" necessarily from the X-15, but it certainly had many generic similarities, such as being air-launched and rocket-propelled. But to have the vehicle reach the edge of space, return to Earth safely, and then do the same thing safely in a short period of time makes SpaceShipOne a real technological achievement.'

Professor Karagozian expresses high hopes for the further potential too. 'If he demonstrates it as a robust and safe system, then there are plenty of people who would enjoy that kind of trip to the edge of space. And, scientifically, there are lots of experiments still to be done in micro-gravity.'

In the aftermath of the *Columbia* space shuttle disaster in February 2003, when the vehicle broke up and burned on re-entry killing all seven of the crew, getting into space regularly has become very difficult. To date, most micro-gravity experiments have been done in drop towers, in which an experiment is released from the top of a very high tower to re-create weightless conditions. NASA has these facilities but the effect only lasts a few seconds.

'*Columbia*'s last mission was to conduct a great many combustion experiments in micro-gravity,' explains Professor Karagozian. 'Many are proposing they could be done on the International Space Station, but SpaceShipTwo could eventually be an alternative micro-gravity test bed. I don't see why not. Of course, Rutan is not the only one working on low-cost space-travel concepts, but no one else is quite as far along in terms of their overall space-tourism viability.'

Maybe she needs to have a word with Jim Benson, who would disagree on this point.

14. GALACTIC EXPANSION

R ockets might be loud and very fast. But for most people they really aren't sexy. If space tourism was ever going to appeal to the masses, it required a makeover. And the ruby-red Virgin brand had the global cachet and marketing edge to make it happen. None of the other players in the commercial space arena had any experience of transporting millions of paying passengers around the globe. And Virgin, operating three separate airlines and carrying over 50 million passengers a year, had an unblemished safety record, never having lost a single passenger in over 21 years of operation. None had the sophisticated marketing machinery of Virgin's consumer-friendly empire nor the high-level networking power of Sir Richard Branson. The Virgin Galactic arrival had created a new supernova.

Virgin's Galactic's snap arrival on the scene changed the landscape and cranked up the volume. It wasn't an unexpected move from Sir Richard. He had already signalled his intentions. In 1999, the company had registered the name

Virgin Galactic as it began discreet inquiries about investing in this emerging sector, and in May 2001, Sir Richard replied to a German journalist on *Welt am Sonntag* who asked him about his personal dreams: 'In the spirit of Dennis Tito we would like to see a time when space travel becomes a real commercial reality through the use of more environmentally friendly reusable vehicles to take more people into space.'

Not everyone was pleased to see Virgin's brash emergence. There were many sceptics inside the aerospace industry – and among the rather po-faced aviation press – who felt it was simply another Branson stunt that had little chance of taking off. But the cynics underestimated the magic of space, and how it would capture the imagination of a new generation of designers and marketers.

Will Whitehorn feels there is a clear distinction between Virgin and the other players. 'Space Adventures are doing fantastic work. But I think that they acknowledge they are not really in the space-tourism business. They are in the private space-exploration business – and doing a fantastic job in getting people onto the big government programmes. Virgin Galactic is different to that, because we are developing a new technology in space . . . it is tailor-made for the customer and the type of customer we are looking for.'

But in the manic days after SpaceShipOne won the X Prize with its new Virgin backer, Virgin Galactic required a top-drawer commercial director who could quickly shape the direction of a new global business that was taking off like a rocket. That person was Stephen Attenborough.

Stephen wasn't really bothered much about space. An Essex boy, born in 1964, he vaguely recalls being brought down early one morning to watch the Apollo missions on the black-and-white television in his family home in Harold Wood. Unlike many others involved in this story, space didn't fire his imagination.

'I was probably part of the first generation just after the Moon landings that grew up with very little real expectation that space had anything to do with me. From when I was old enough to know anything, NASA went into its decline and the public interest had waned,' he says. 'But now I am personally very excited about the prospects.'

He spent his early career in financial services, starting as a fast-track graduate with the UK's major insurance company, General Accident. He had a spell as a broker before joining the fledgling Gartmore Investment Management in 1989. Here he prospered, rising to become one of the thirty senior equity-holders in the business, finally becoming head of the firm's global consulting group. As a founder member, he helped build and shape Gartmore from scratch to a £40 bn-plus firm in the pension market.

But when he reached forty Stephen decided it was time to do something fresh and took six months off to recharge his batteries. 'I had a successful career but, like lots of City jobs, things change, and it was the right time for me to move on. It had been a great place to be.'

This was when the Wadhurst connection kicked in – not quite a mafia, more a close-knit network of friends. Stephen knew Will Whitehorn as they both lived in the picturesque East Sussex village, an hour by train from central London. They shared a mutual friend called Neil McIndo, an old university pal of Will, who had worked with Stephen.

On the Saturday evening immediately after the X Prize was won, Will was visiting Stephen's home for supper. He showed off the DVD of the winning flight and became extremely animated. By 2 a.m. an exuberant Whitehorn was still talking excitedly about its significance.

'We virtually had to kick him out,' recalls Stephen.

Stephen had taken the opportunity of his sabbatical to buy a dilapidated cottage on the south coast, at Winchelsea, near Rye. It was a change of scene and he was rebuilding the house. 'I was up a ladder painting the top of the house and my phone went. It was Will and he said, "Look, do you just fancy coming to help us out for a couple of months or so to set up the space hotline?"'

So he cleaned up his brushes, put on a fresh shirt and went straight up to London to meet Will, Alex Tai and Jonathan Peachey at Virgin Management in Campden Hill Road.

Over the previous weekend, Alex – who was still flying as a captain for Virgin Atlantic flights – had set up a website and a rudimentary database was created. But already the emails were piling in and the phone was ringing. There were loads of potential clients, with people clamouring to buy tickets – but no business infrastructure.

Will's view was that Virgin Galactic didn't need a space nut but somebody with experience of starting up a new organisation and putting the fundamental bricks and mortar in place. But what was paramount was how to treat the clients, who were the future of the operation. They all needed to get the Virgin five-star service.

'I jumped at it,' says Stephen. His wife, Amanda, agreed, although she knew Stephen had two blue-chip investment job offers with lucrative packages that were far more secure for the family. 'But Will's offer of a few months' work until January 2005 was quite appealing. I hadn't really found anything else that I wanted to do. But this sounded fantastic and I thought I would give it a go.'

The following Monday, Stephen walked through the door at Virgin Management's office and instantly found the culture fitted him like a glove. 'It was so open and welcoming. I also felt I had a lot to contribute. There were plenty of areas where my skills were transferable. It was a revelation.'

The launch of Virgin Galactic had been a last-minute affair. It all hinged around whether the deal could get done with Paul Allen and Burt Rutan – and whether SpaceShipOne was going to win the X Prize.

Stephen's job was to prove that Virgin Galactic could work from a market perspective. He needed to find out if the demand was real – and not a chimera. 'It was about setting a realistic price and ensuring we get paying passengers on board. We were selling a dream to high-net-worth individuals and demanding customers. We were selling something that has an uncertain future, because it hasn't been flying commercially yet. And, funnily enough, the management of relationships and expectations is actually very similar to the hedge-fund industry where I worked.'

He set about creating a reservation structure for paying passengers, working particularly on the terms and conditions. In the opening week, he sent a note around introducing the concept of Founder – the price for Founders would be set at $200,000 – and Pioneer astronauts (later the class of Voyager astronauts was added). It was an idea that would develop arms and legs.

What became apparent was that some people were quite desperate to go into space. Attenborough felt it was important to ensure that everyone was clear about what exactly was on offer from Virgin Galactic. This was a suborbital trip into space – and something as simple as this had to be precisely explained. And there would be a premium to be paid for an early flight. The obvious first thing was to find out who wanted to fly as soon as possible, so Stephen set up a booking and a deposit structure. But it was not without early headaches.

One critical issue emerged early in the process, which Stephen handled with tact and diplomacy. Bill Cullen, the chairman of Renault Ireland, who is an author and a well-known celebrity business figure in the Republic of Ireland, was one of the first to register. He knew Sir Richard Branson and had been discussing some business deal together and said he was keen to fly. He had emailed very early on asking for a seat.

'I'm going to come to London with my cheque for $200,000,' he said, when Stephen phoned him to have a chat.

'Well, we're taking reservations, but we aren't collecting money as yet.'

'I don't care. I want to give you a cheque.' So he flew across from Dublin and handed over the money.

'He was our first one and we gave him a letter as a receipt. We didn't cash the cheque though, we put it in the safe,' says Stephen.

A few days later, there was a call from a rather perturbed Irishman, Tom Higgins. He had heard that Bill Cullen was saying he was going to be the first Irishman in space. Tom was upset. 'I'm pretty sure that I was the first one to register. I think my name would have registered before Bill Cullen and I want to be the first Irishman in space.'

Stephen Attenborough had a delicate problem to deal with. There was no way he wanted to lose these big-hitting Irish fellows.

'I said, "Well, the trouble is you might well have registered before but Bill has paid and I'm not really in a position to say who will be first."'

Suddenly there was an issue. 'I had been assuming, up until that point, that for Founders we would probably just allocate on a first-come, first-served basis, but it quickly became apparent that this was not going to be a solution that would work.'

It was perplexing for Attenborough, who had a nightmare of a revolt on his hands. Then one night, lying in bed at his East Sussex home, he came up with a solution. 'We had already decided to restrict the number of Founders to one hundred, and to collect the money in a dignified period and make sure the bookings were done properly, and that everybody got the chance to buy one. Then I thought, let's just put everybody into a draw, so that there was an equal chance of everybody being on that first flight.'

Stephen spoke tentatively to the early Founders who had already signed up, asking them how they felt – they all came back and agreed. For Tom and Bill, it wasn't a perfect solution, but at least they are on an equal footing. And it saved a lot of red faces too.

'It became very clear to me in talking to people that there was a distinct difference between my previous hedge-fund clients and our early Virgin Galactic customers. The difference was that Virgin Galactic's Founding astronauts wanted to be an integral part of the evolving project; this was something seriously important to them. To almost all of them, this was far more than a two-hour jaunt into space. It was also a stepping stone for the future – and they wanted a share in this. People put a very high price on flying early – and a price on getting special access to the project. The Founders are quickly becoming an important part of the wider team – their input helped us shape the experience.'

This flood of money was to be used in the space business – with a guarantee that until Virgin Galactic is able to confirm flight dates, Founders would get their money back in full. 'We took a big, deep breath before deciding this. It's not a typical thing to do that – but it proved our commitment,' says Attenborough.

An informal 'Space Team' was set up early on to look after the company. One influential addition was Ned Abel Smith, a smart and eager creative-marketing type in his mid-twenties. He was Sir Richard's nephew and had set up his own business in his early twenties.

'He has been hard-working and brilliant. He is almost proprietorial about Virgin Galactic,' says Stephen.

At the Virgin Christmas party in December 2004, Stephen was sitting at a table with Will Whitehorn. They were both delighted with how much had been achieved in such a short time. Stephen said he needed to make a decision on whether to stay or go by 14 January. Will was candid. 'We would like you to stay here for another six months with the likelihood that a full-time position will be extended from there. But I can't guarantee this. It's not an easy decision – but if I were you I'd take it,' he said.

Stephen still had two job offers in his writing bureau at home. 'Then I got a call from Richard in Holland Park to go and see him. It was the first time I'd actually met Sir Richard. He repeated what Will had been saying, adding, "I'm going to make this work – and it would be great to have you on board." What could I do? I was almost certain,' says Stephen.

Then a leaked diary item in a Sunday newspaper made Stephen burst out with laughter. 'Here was a story in the newspaper saying that I had already joined Virgin Galactic.' The snippet was upbeat but sarcastic. It said the only difference for Stephen Attenborough was that in his last job he had to warn clients that investments *could* fall as well as rise, whereas in this new job it was a promise.

Stephen was hooked – he was now on board. But Virgin Galactic was expanding and there wasn't enough room at Campden Hill Road, so some plush Virgin offices around

London were surveyed. Then there was Half Moon Street – a highly appropriate address for a suborbital travel business.

On a cold February evening, Alex Tai and Stephen picked up the keys from the agent and arrived at an anonymous front door in Mayfair. There was no electricity as they entered, so they used the lights on their mobile phones to negotiate the narrow stairs. It was a warren of rooms in a decrepit Victorian townhouse, opposite the old Army and Navy Club. The pair stumbled around the derelict offices that had been used for a score of long-forgotten Virgin projects over the years. The pair adjourned for a beer in nearby Shepherd's Market to discuss the place. It was hardly Mission Control, Houston. It was brass monkeys in winter, sizzling in summer and probably haunted. They agreed it was just the right location to start Virgin Galactic.

Alex was away flying on the first day at the new office, so Stephen Attenborough sat alone with a box of Virgin stationery. It was just like a normal start-up company – arranging phone lines and computers, setting up an Internet connection, etc. – but with one major advantage. It had the might of the Virgin brand four-square behind it.

One event was an eye-opener for Stephen about how true this was. A few days after moving in, Arpad 'Arky' Busson, the former partner of supermodel Elle MacPherson, contacted Sir Richard asking him to support a charity event. Arky is a successful hedge-fund financier who set up his company, EIM, in Switzerland in 1992. Its London office is now in Mayfair, near Half Moon Street. He is one of Europe's wealthiest financiers.

Charismatic Arky also runs a philanthropic charity, supported by the hedge-fund community, called ARK – Absolute Return for Kids. Each year ARK holds a major gala event where they raise an extraordinary amount of money through auctions. Children in hundreds of projects across Europe and Africa have benefited from the proceeds of this glittering London occasion.

'I didn't know Arky but I had met him briefly before and I knew of the reputation of ARK and the auction. They wanted a space flight – and previously we had refused every other approach. My view was we should try and do a deal that kept him – and us – happy. We set some parameters as to what we should do in order to ensure everything was in order.'

There were some quite hard negotiations, but the deal was finally sanctioned by Sir Richard on the day of the auction. 'I then had one of these bizarre Virgin Galactic moments when I was sitting in the black limousine of Stanley Fink, the chief executive officer of Man Investments, the largest listed hedge-fund company in the world, with Sir Harry Dalmeny, the Sotheby's auctioneer. He was whizzing around London picking up his dinner suit while I was briefing him about Virgin Galactic so he could auction it.'

Later that evening Stephen got a text message from Harry saying it had been an outstanding success. The wealthy revellers had made the highest bid they had ever had at an ARK auction. The space flight had gone to a guy willing to pay $500,000, which was more than double the list price. The auction itself raised more than £11 m ($21 m).

'It was great for the charity – but it was also great for us as a proof of concept. We had always had it in mind to do something and this proved it worked.'

It was becoming apparent that Virgin Galactic was a very valuable name – and that made it interesting for other brands. Doing deals with other global companies would become a major source of revenue for what was a very expensive programme. But product placement was another opportunity that astounded Attenborough.

'I got a call from Richard in Australia. It was very early in his morning and he'd obviously had a very good night. He said he'd just had dinner with a fabulous guy called Brian Singer.'

Brian was a hotshot young film director. He was directing the new Superman film *Superman Returns* and he wanted to go into space. Super-salesman Branson persuaded him to make a reservation. 'We were able to book him up and he has become an enthusiast Founder,' says Stephen.

But Brian's people came along with another idea involving the film, which has an amazing space scene in it. The movie is set in the future and portrays the launch of a new passenger shuttle service attached to the top of a plane. Lex Luther typically tries to destroy it before Superman flies in to save the day. Brian Singer suggested it would fun to use the Virgin Galactic logo on the spaceship.

'We said, "Let's have a quick look at the script, to see if it fits with what we're trying to do."' The last thing Virgin Galactic needed was any association with a disaster movie. But the final script was still top secret. A senior film executive, with the script in a locked briefcase, was dispatched from Warner Brothers in LA to Half Moon Street to reassure the Virgin Galactic space team that it would be great exposure – having Superman save the day.

'We loved it and we said yes,' says Stephen.

Then the producers came back with yet another scheme – offering Sir Richard and his son, Sam, cameo parts in the film being made in Australia. Here was another offbeat idea that suited the company. It's really a Hitchcock moment. 'If you look very closely you can see them. It's a bit of an in-joke, though,' he adds.

The numbers on the Virgin Galactic database just kept growing. Attenborough would spend the evening at home with a bottle of wine sifting through the thousands of names. One night he spotted an email from a woman called Victoria Principal, who was a major soap star in the 1970s and 1980s with the hit series *Dallas*. 'I saw the name and I thought, I wonder if that's the television star. I just phoned her up and it was. She's lovely.'

Victoria Principal recalls returning Attenborough's phone call and bombarding him with questions about how the spaceship would operate.

'Stephen said to me that I was the first person to ask so many technical questions about the fuel, the duration and the impact on the environment. I grilled him about all the green issues. Then I spoke to Richard about it too.'

In general though, there was little interference from Sir Richard except one early word of advice to Stephen. 'By the way, I think it is very important that there are no freebies – and no upgrades.'

'I didn't have a problem with this,' says Stephen. 'If people are really passionate and fortunate enough to have the wherewithal they will want to buy a ticket. It doesn't really matter if they are a celebrity or not. We have signed some celebrities up but we have signed up lots of ordinary people who have paid their money – they all share the same factors.'

But he has been surprised by the number of celebrity agents who asked for a free flight. Without specifying names, he says some have called up demanding to be on Virgin Galactic flights by virtue of their supposed star status. 'We have said "no" to all of them.'

It dawned on Attenborough early on that there was much more to the commercial space project than simply giving a relatively few rich and fortunate people a fabulous ride. He believes it is important to be able to justify spending this sort of money for five minutes of weightlessness.

'There are a lot of pressing needs in the world, and many environmental concerns, but this has been at the forefront of our thinking. One of the reasons Virgin Galactic decided to go into this is the belief that we need to be investing in breakthrough technologies. A growing part of this is looking at technologies that will be environmentally sustainable in the future.'

Attenborough is inspired by the prospects ahead. He genuinely believes Virgin Galactic and the first passengers who have signed up are the trailblazers in the future of commercial space travel. A lofty ambition.

15. THE NEW MEXICAN WAVE

Mike Foale was acting the clown – and the kids were loving it. The shuttle astronaut was standing in a marquee that might as well have been a Big Top – because 400 children were screaming with laughter, their sides splitting with mirth. On a giant screen in front of the classes of eleven-year-olds from elementary schools in Las Cruces and El Paso, Foale was shown swimming right through the International Space Station like Superman. Arms extended, he surged towards the camera like a superhero. The kids lapped it up.

Then he played with four coloured sweets hanging in zero gravity. 'NASA doesn't allow us to advertise, so they are described as chocolate-coated candies,' he told them. They all knew they were peanut M&Ms. Foale juggled with the little balls and then gobbled them down one by one in front of the kids, adding that he doesn't eat vegetables in space because they give him too much wind. They all guffawed again. He told them how, in space, you always eat with your mouth firmly closed – otherwise the debris gets everywhere. His final trick was blowing a water bubble, pricking it with a very fine straw and blowing in some more water. It was wonderful and surreal and hugely entertaining.

Foale is a born communicator – he has been able to sell the sizzle about space and inspire young people. But he was the warm-up act for the next visitors. The kids were still shouting excitedly when they were asked to welcome Bill Richardson, the Governor of New Mexico, Anousheh Ansari and Buzz Aldrin. Seldom had Richardson, a wily political operator, had such an enthusiastic greeting. This was the main political event at the Wirefly X Prize Cup in Las Cruces in October 2006.

'OK,' began Richardson. 'How many kids are here from El Paso?'

Loud cheers from the left.

'And how many kids from Las Cruces?'

Even louder shrieks and screams from the right.

'First, let me say that Peter Diamandis is the originator of the X Prize Cup and I want to thank Peter for giving me credit but it's Peter who chose New Mexico, so let's here it for Peter.'

More cheers and clapping.

Then Buzz Aldrin took a bow, to uncontrolled applause.

Richardson paused to make a serious point: 'The Space Age has come to New Mexico.'

So how did New Mexico manage to attract the space elite? In late 2001, Rick Homans joined up with Bill Richardson. Richardson was a Congressman and ambassador to the United Nations and then energy secretary under President Bill Clinton. Homans was brought on board to help him run for governor in New Mexico.

Homans started out as a newspaper reporter before becoming an editor and publisher, but was always involved in politics. Indeed, he ran unsuccessfully for mayor of Albuquerque in 1995. Richardson asked him to develop all the policy for the campaign and the administration that followed. The goal was to have clear and realistic plans in all areas, from health to economic development, so they could hit the ground running after the election. When Richardson was elected he rewarded Homans with a powerful position.

'When I started my state government job in January 2003 as the cabinet secretary for economic development, there was absolutely no way that I could have predicted that I would be in charge of building the world's first purpose-built commercial spaceport.'

New Mexico has a history of lagging behind the rest of the United States. In the years before Richardson's election economic development hadn't been a priority. Richardson had a lot to do – and he started on day one. He cut personal income taxes by 40 per cent, he cut capital gains tax by 50 per cent, and created new tax incentives for companies to come to New Mexico. More interestingly for space tourism, he allowed the state to take a direct equity stake in new companies. It was all part of a strategy to create more opportunities for New Mexicans.

Richardson and Homans wanted to get in on the ground floor of brand-new industries with major potential for growth – New Mexico is not short of natural assets and advantages. The governor was keen to build on the aerospace and nanotechnology industries in the state and saw opportunities for biotechnology and renewable energy. Aerospace was a natural for several reasons. One reason was Vern Raburn, the entrepreneur who created Eclipse Aviation in Albuquerque in 1998.

Raburn was a hugely successful hi-tech entrepreneur who worked with Microsoft, Lotus Development and Symantec, but his enduring passion was airplanes. In the mid-1990s he saw the growing market for a high-quality, low-cost private jet. He became committed to lightweight engines after a NASA initiative was set up to look at how hi-tech could transform the aviation business.

Raburn was joined by Dr Oliver Masefield – an aircraft designer with nearly thirty year's experience – and they set up the Eclipse 500 development team. In 2006, it began producing its first planes.

'We're very proud to have Vern in Albuquerque. Eclipse has

invented the first lightweight business jet that will revolution-ise the aviation market,' says Homans.

Another big positive for New Mexico was that White Sands Missile Range was the proving ground in 1947 for the US's space programme, when Wernher von Braun fired his rocket into space, while Robert Goddard experimented with rockets in the south of the state. Then the White Sands Test Facility was set up to perform testing of vital flight systems for the Apollo mission to land men on the Moon. Construction began in 1962, and the first tests were performed in 1964. All the Apollo astronauts lived and worked in New Mexico.

Today NASA's White Sands Test Facility is a world-leading site near Las Cruces, New Mexico. It tests and evaluates spacecraft materials, components and propulsion systems and is still a training ground for space shuttle crew. The remote site, covering over 28 square miles, allows engineers and technicians to test hazardous materials and components. And it will be close enough to the new Spaceport to become a vital part of space tourism's future.

Rocket engines with up to 25,000 lb of thrust under near-vacuum conditions – similar to those experienced in space – can be fired up here. And it also provides chemical and metallurgical support that will be vital for servicing a new generation of commercial spacecraft. Many of the younger entrepreneurial space businesses have been working out of tiny workshops, garages and university labs, so White Sands will given them a safer place to try out new designs. And the White Sands business people have become aggressive in wooing the next generation of space operators, especially for depot maintenance and repairs. Contractors involved with NASA and other programmes can make arrangements directly to use the facilities – which will be another boost for New Mexico.

Homans points to another connection with the state – Roswell, the home of the extra-terrestrials. Roswell is the site of one of the most famous intergalactic visitations, and the town's famous UFO Museum is also a research centre for the study of alien sightings. Some have already suggested that Spaceport America is just a plot by aliens to get back home.

Building on all this history, New Mexico has become an alternative landing site for the space shuttle, with the US Air Force research laboratory's Space Vehicles Directorate at Kirtland and the NASA facilities at White Sands being vital for the space station and the space shuttle.

This spurred a group of dedicated space enthusiasts in southern New Mexico to set up the Southwest Space Task Force, led by Donna Nelson, in the early 1990s. The goal was to position the state for the next generation of renewable-launch space travel. The founders of the task force – now renamed the New Mexico Space Alliance – were committed to pressing their vision of a spaceport on Richardson.

'Within days of starting my job in January 2003,' says Homans, 'this group came to me and wanted to explain why the office of space commercialisation was within my department. They explained why a specific location in New Mexico fit the bill.

'The conversation with the task force seemed a little far-fetched at the time, but they convinced me that we had an extraordinary opportunity. I also understood that we would have to wait for exactly the right opportunity and customers to come along.'

Homans could see that the task force had some powerful arguments for the spaceport. 'Firstly, it was adjacent to the White Sands Missile Range, which has restricted airspace from the ground to infinity. This created a natural pathway into orbit . . . the only other place in the United States that has that same configuration of ground to infinity is over the White House. It is an unusual and rare piece of air space in the US.

'Secondly, the elevation of the site is 4,700 feet. So for vertical rockets going into space it lowers the cost and the need for fuel and increases the payload capacity compared to sea-level launch sites,' says Homans. 'Third, the area around the site is sparsely populated. If you fire a rocket and you kill a cow, you have to buy it. That's actually a law. You can eat it, but you have to buy it.'

So this area of New Mexico, with its low density of people, lowers the insurance premiums and the risks. This makes it much easier to gain launch licences from the FAA.

'Fourth, we have exceptional weather in New Mexico. It's clear and bright every day. When you launch a rocket every day you don't want to rely on prayers and good luck. You need a reliable launch schedule, otherwise it begins to cost a lot of money.

'Fifthly, we have the right air. If you put equipment on a coastal site it will be subject to the destructive forces of corrosion within months or a year, yet that same piece of equipment lasts ten years or more in the dry desert.'

All of this impacts on the bottom line of a commercial launch vehicle operation.

'They also told me there had been numerous committees, studies and reports focused on Spaceport America, and they had all offered two conclusions. First, the spaceport would probably cost $200 to $300 million to build because of the extensive runways that would be required. Second, that we should not follow the "build it and they will come" philosophy. Instead, we should wait to build until the new industry arrived and, more importantly, we had an anchor tenant knocking on our door.'

Homans thanked the task force for the information. But he had more pressing matters to attend to and Richardson was still a new boy in the state house. It was put on the back burner – to wait and see what might happen. A few months later a package arrived on Homans' desk from a group called the X Prize Foundation. He flicked through the document and a light bulb came on in his head. He saw that 24 companies from 7 countries would be competing for this $10 million prize.

'I couldn't even pronounce the name of the guy who had put all this together, but I understood that he was looking for a place to hold an annual competition once the X Prize – which took place in the Mojave Desert – was over.' This would become the Wirefly X Prize Cup, set up so the privately funded space companies could compete against each other. This was one of the first real opportunities for the spaceport and New Mexico. 'It was a do-or-die decision. If New Mexico was serious about the emerging reusable launch-vehicle industry, we simply had to bring the X Prize Cup to the state,' adds Homans.

The 24 companies were a mixed bag of businesses. Some were living hand to mouth, some backed by successful entrepreneurs; they represented the chaotic beginning of this new industry. These were not the Boeings, Lockheed Martins and BAE Systems.

'A new industry at the beginning doesn't look pretty and there is a lot of failure. But in economic-development terms that was what we needed to grab hold of,' Homans explains.

If the X Prize Cup had gone to another state, it would have given the birthright to another part of America. The politicians recognised that going after this would send a strong message about the brand of New Mexico. It would signal that the state was entrepreneurial and adventuresome, that they were willing to take risks to embrace innovation and next-generation technology. Homans' team put in a strong bid with $9 m of state funding plus infrastructure support, and New Mexico won the X Prize Cup beating competition from the bigger states of Florida, Oklahoma and – always a bonus for New Mexicans – neighbouring Texas.

In early 2004, the X Prize Cup was announced, which made international news and branded New Mexico as one of the launch pads for the new technologies. It was an industry in its infancy, but the Cup would quickly become the must-be-there gathering place for all those serious about the industry.

One of New Mexico's first supporters was an ambitious rocket man from Manchester. Steven Bennett, who was out at Las Cruces for the X Prize Cup in October 2006, has the indomitable British bulldog spirit. He is a true Brit who wants to rekindle the days of the British rocket industry, which was left to wither on the vine by the UK government in the early 1970s. He admits he watched too many episodes of *Thunderbirds* when he was a kid and decided to become a rocketeer in 1992 after a career as a chemist with Colgate-Palmolive. After years of effort, in 1998 he set up Starchaser, a company based out of Salford University. Its learning outreach has been pivotal in teaching children about rockets in the UK. Then, in

January 2005, he moved some of his operations to New Mexico, buying 120 acres of ranch land just off the Interstate 10, where he is now building a factory, an astronaut training centre and a space theme park.

'We were the first private space company to come to New Mexico, we were here long before Virgin Galactic and even before the X Prize had chosen this site. It's got the great weather, it's got the elevation and the de-listed airspace next to it. I said, I'm going to be here.'

Bennett was also preparing for a 50,000 lb LOX rocket launch at RAF Spadeadam in Cumbria, the site built to test Britain's Blue Streak and Black Arrow rockets in the 1960s. He was keen to breathe life into a forgotten part of Britain's technological heritage. 'I wanted to build a rocket that would take me into space.

'I'm quite a competitive person. What I want to do is go higher than SpaceShipOne and I want Starchaser to be the first company to carry fare-paying passengers into space. I've got people already lined up for the first flight – we've sold tickets for £250,000 each.'

He said the people who bought them said to him, 'we don't care how much they cost' as long as we can go into space and come back safely. 'We want to beat Virgin Galactic with fare-paying passengers and we want to go higher – and I think we can do it. I don't think SpaceShipTwo is going to be ready until 2010. They may do it sooner, and good luck to them. What we are doing is a completely different approach. Ours is a ballistic flight. When you fly with us you get to experience the Alan Shepard and Gus Grissom flight. You do the training, you put a spacesuit on – you're in the capsule and you do everything that they did on these first historic flights. You truly will earn your astronaut's wings.'

The Spadeadam tests are the fourth in a series of major tests put together by Bennett and his team. He started his programme in 2000 with a 500 kg thrust engine, the patriotically named Churchill Mark 1. The Churchill Mark II followed, which was six times more powerful, then the heavier Mark III was built and tested. But they have stepped back with Storm, which Bennett predicts will launch their Skybolt sounding rocket into space.

So, as Homans was welcoming Starchaser, he also prepared legislation for the state's 2005 session, putting up a million dollars to create the independent New Mexico Spaceport Authority, which would run the facility like an airport, issuing bonds and selling leases. Homans met with Patty Gray Smith and her staff at the US Federal Aviation Authority to get the spaceport licence approved as quickly as possible.

On 4 October 2004, the industry had arrived in the form of SpaceShipOne, WhiteKnight and a unique cast of colourful characters, including Brian Binnie and Mike Melvill. But Homans also heard something else. 'We also heard, loud and clear, Sir Richard Branson's announcement about Virgin Galactic. I told my staff that day that they would have to bring Virgin to New Mexico because they were credible and they were leading the pack. Virgin would be the anchor tenant for the other condition we needed before we could build our spaceport ... one cannot underestimate the significance of Branson's announcement,' says Homans.

'Here was one of the world's richest men, one of the most respected entrepreneurs and a trailblazer in the aviation industry, going to build a business around taking people into space. Suddenly, the world had changed and, in New Mexico, we felt we might just be in the right place at the right time,' he adds.

In February 2005, the New Mexico state legislature committed $110 m and expected to receive the other $25 m for road construction. In the spring of 2005, Homans had to prepare a solid case for the economic impact of the spaceport. He asked New Mexico State University to assess the industry and forecast the potential. As those results came together, Homans also decided it would be prudent to get internationally respected analysts to conduct some research on the forecasts. He approached a space-tourism consultancy, Futron, who had built a reputation for assessing the tourist potential of different destinations. After all, outer space would now be competing with other places for the world's leisure spenders.

When he received the findings, Homans was pleasantly surprised. The numbers were stacking up. The New Mexico State University survey showed that the spaceport, after five years, would sustain about $1 bn in new revenues, with $350 m in new payroll and 2,800 new jobs. Futron looked a little further down the road and suggested that New Mexico might generate up to $750 m in new revenue, and 5,800 new jobs by 2020.

This was the arrival of a major new industry, involving research, testing and operations at the spaceport. There would also be the spin-off of thousands of tourists just coming to visit the spaceport to get a small taste of the action and to watch the shuttle flights into space. Further, it was anticipated there would be some manufacturing and supplying of parts and specialist maintenance.

'If we are right about what we are seeing happening over the next ten to twenty years, with the breakthroughs in technology and the increased access to space and the development of space vehicles, then there are a lot of new jobs at stake,' he states.

For California or New York, the number of jobs would be a drop in the ocean, but in a state like New Mexico, this would transform the entire economics of the place. 'Politically, this has been something easy to get everyone to rally around, while it would be much more difficult in a larger American state,' admits Homans. 'If you look at these numbers, the impact for New Mexico would be big, really big.'

The reports were critical in setting out the case to the New Mexico legislature, the federal government and, more importantly, the local communities. Richardson and Homans then had to go for the hard sell. But they also had to make it clear that this was a risk.

Meantime, Homans' attention was also on delivering the X Prize Cup to Peter Diamandis and his supporters. A stunning preliminary event was created, called the Countdown to the Cup. In October 2005, 15,000 people turned up to the show to hear and feel the static displays and rocket launches. The event featured a live rocket engine firing and the first demonstration of XCOR's E-Z Rocket, flown by space shuttle astronaut Rick Searfoss, plus a demonstration by Armadillo Aerospace of their vertical takeoff vehicle. XCOR was also working with Peter Diamandis in setting up the Rocket Racing League, launched in October 2006, in New Mexico, and using rocket-propelled aircraft to race around a track in the sky defined by a series of digital gates visible on the pilot's cockpit display.

Winds picked up and two giant 30 ft inflated balloons – one of Earth and one of Mars – began to break loose from their moorings. Ned Abel Smith from Virgin Galactic was holding onto Earth and it started rolling over and going down the Tarmac. He yelled down his radio, 'Earth has broken loose, but I'm going to save Mars!' It was that kind of haphazard happening.

In the interim, Homans had begun serious negotiations with Virgin Galactic. 'We started to make contact with the team from Virgin Galactic. Representatives from New Mexico started to meet in Washington, then in London.' Homans and his team were now getting regular visits from Will Whitehorn and Alex Tai, but every time they turned up in New Mexico it would either snow or rain. He joked that his credibility about warm balmy days and clear blue skies was beginning to suffer.

Rain showers aside, Virgin Galactic could see New Mexico was the ideal platform for long-term growth, frequency of flights and the planning of a new and exciting industry. Much of the state began to rally around the ideas too. On 14 December 2005, Sir Richard Branson and Governor Richardson signed an agreement for Virgin Galactic to have its world headquarters in New Mexico, alongside was the former *Dallas* actress Victoria Principal, now an ambassador for the Virgin space project.

Not everyone was so welcoming. One critic of Richardson, John Grubesic, complained that the governor had vetoed a proposal to improve Santa Fe's County Judicial Complex and cut funds to a paediatric oncology facility while supporting

Virgin Galactic. But a clear majority recognised the benefits. The state agreed to build a permanent facility at the spaceport for Virgin and the company signed a twenty-year lease that would pay for the construction and financing of their own facilities over the term of the lease. The deal was struck and for the first five years the lease payments were to be capped at no more than £1 m per year. Virgin Galactic would also pay user fees at the spaceport.

The editorial in the Santa Fe paper, the *New Mexican*, welcomed the news – with a proviso.

The news that New Mexico and Virgin Galactic are teaming up to build a spaceport – that's right, a spaceport – in southern New Mexico is a bit amazing. It's Buck Rogers come to life, right here in our back yard. The spaceport in the desert is the brainchild of British entrepreneur Sir Richard Branson, whose Virgin brand started life as a record label, then segued into a travel and airline business. The visionary billionaire decided that space tourism was the next big thing and needed a partner to move his venture forward.

This is uncharted territory, and we're pleased that Bill Richardson's deal includes milestones that Virgin must meet before all of the state money is turned over, a safeguard that protects the state in case the project turns sour. We also like the idea put forth by a former US Senator. Harrison Schmitt, a former astronaut, says that the state should make independently sure the spaceport will fly before pouring taxpayer dollars into it. In any project this grand, it's essential, too, to make it as environmentally sensitive as possible.

Homans admits there were fundamental issues to tackle. He agreed it would be a tough task trying to justify the expense of a $225 m spaceport in a relatively poor state of only 1.8 million people. 'The plan relies on a number of funds; $135 million will be coming from the state government and we expect the remaining $90 million from a combination of federal funds and local communities, where there will be a small tax that will help build the spaceport.'

But in New Mexico there is a legend that still haunts many people in the know. It involves Bill Gates in the early years of Microsoft. Most people do not realise that Microsoft was set up in Albuquerque in New Mexico. Gates was short of cash and a little rough around the edges at the time. He got in trouble for some unknown reason in the mid-1970s (the police records have all been expunged from the police department, but a mug-shot still exists of a recalcitrant Gates). It was a different era and Microsoft, at the time, only employed about seven people. Everyone wore sandals, cheesecloth shirts and didn't take showers every day. Gates made the rounds of law firms and banks looking for loans. He even approached a law firm and asked them if they would trade stock for some cash. Just like the record company who rejected the Beatles, they missed a golden opportunity. Gates was shown the door and eventually he moved back to Seattle.

'What Gates left behind in New Mexico was a profound lesson to us,' says Homans. 'He has always been very kind and doesn't point the finger at his rejection in New Mexico but I feel very strongly that if New Mexico had developed a different business culture – a culture which embraced entre- preneurs and new technology – then Bill Gates might have stayed. The history of our state would be very different.'

For Homans, the spaceport and the Virgin Galactic deal is a chance to learn from the missed opportunity. 'It is imposs- ible to forecast the impact some development will have on our lives and on the world. I suspect that the second space age will be equally profound. Peter Diamandis, Sir Richard Branson, Burt Rutan, Paul Allen and others have said their goal is to make space accessible and affordable and once that happens our lives will change.'

The state of New Mexico has now signed a contract with a design and engineering firm, and the ground at the $225 m world's first purpose-built spaceport was broken in early 2007. Architects and designers from Spaceport America have been to London for the full Virgin experience. They were taken to the Virgin Clubhouse for Upper Class passengers at Heathrow airport, and were briefed about Necker Island and Morocco, both exclusive luxury destinations for discerning and rich tourists. And they met Philippe Starck, the famous designer who has put his visionary skills to good use for Virgin Galactic. They should be under no illusion; the spaceport is to be an eighth wonder of the world, in terms of design and customer experience. It must be tasteful and environmentally coherent too.

'By 2010, if all goes well, we'll be watching Virgin Galactic make regularly scheduled flights to space from New Mexico. By that time, we look to have other tenants as well – some of which we expect will be flying cargo and passengers to the International Space Station and, soon thereafter, the Moon. This is history in the making and every step forward of this project feels really thrilling – to have the opportunity to be in the centre of a historical moment,' Homans says.

Homans has never pretended to be a space buff but he does remember the Eagle landing. 'I was thirteen when Neil Armstrong and Buzz Aldrin made their first historic steps. I do remember the awe and the excitement. But it didn't change my life the way it did for many who went into the space industry.' He admits he is a relative newcomer to the space-tourism industry and he had to be convinced about the significance of the X Prize, of SpaceShipOne, Virgin Galactic and of building a spaceport in the desert. But a converted sceptic is the best kind of born-again zealot. Homans says the world is heading for a second space age. 'I became convinced of that point when I proposed to Bill Richardson ... that he persuade his political colleagues to build a $225 million spaceport.'

Richardson and Homans worked together to convince 144 legislators in New Mexico to back the plan and put up the hard money. 'When Will Whitehorn was there with Peter Diamandis presenting to the New Mexico legislature in the hall, one of the senators who does not know a great deal about space travel said, "What if one of those whackos gets into space and just keeps on going? What are we going to do about that?" Whitehorn couldn't be kept in his seat. He jumped out from behind the podium, interrupted the session and started to engage the senator in the hall.

'Which is not the way you do it,' says Homans. 'In a state legislature, you respectfully nod your head and say: "Well, that may be a problem, but we think we have it under control."' Homans points out that was the last time he had Whitehorn presenting his case.

The vote taken by the legislature in February 2006 marked a turning point for the state. Homans appreciates there is a huge challenge for the fledgling commercial space industry.

'I think it's a good story about how New Mexico became a believer and how building a spaceport in the desert became a good investment. In the past, I have represented the primary challenge that this new space industry has: becoming credible to the government bureaucrats, venture capitalists, bankers and potential customers who will pay the bills. I am a relative newcomer to your industry – and a person who, in the beginning, had to be convinced of the significance of the X Prize, of SpaceShipOne, of Virgin Galactic. In time, I came to believe that we are on the threshold of the second Space Age.'

There was one pressing matter to settle – leasing the land from the ranchers. On 21 December 2006, the New Mexico Spaceport Authority finally secured a historic long-term agreement for access to the ranch land for Spaceport America.

At a session in Truth Or Consequences, separate legal agreements were signed between the spaceport, the state's land office, Sierra County, and the two private ranch operations. 'The agreements provide the ranchers and the State Land Office fair compensation for the land, and they make it possible for us to proceed with planning for Spaceport America,' said Homans. 'The agreements also protect the ranching heritage of this beautiful valley, and prohibit any further mineral exploration, road development or commercial development that could impair the state's ability to operate a commercial spaceport.' If it is determined that future spaceport operations damage the ability of the ranchers to operate commercially, they will be entitled to compensation – or even to sell out completely.

New Mexico's commissioner Patrick Lyons made it a condition of negotiations that the Spaceport Authority had to reach agreement with the ranchers before he would enter into a separate business lease for Spaceport America. Now this has the go-ahead with a 25-year lease, and an option to renew for successive 25-year terms, at an initial rent of $25,000 a year.

New Mexico's leadership has also inspired the Swedes to approach Virgin Galactic to build a spaceport in northern Sweden, where tourists could fly up to see the Aurora Borealis.

Dr Olle Norberg, head of Esrange, the Swedish Space Corporation's missile and rocket range station near Kiruna, was signing a deal with Virgin Galactic in 2007. Kiruna is a town of 22,000 people and home to the largest underground ore mine in Europe. But its remote location was also ideal for the Swedish Space Corporation.

'The Esrange has been there for forty years and we've launched five hundred and fifty Saturn rockets. So we have the experience to fly rockets and to handle these kind of propellants and vehicles with safety,' Norberg says. 'It will be a different experience from New Mexico and complementary in many ways. The tourist can come and stay at the Ice Hotel and go out into the icy wilderness and can then take a trip up into space, and look at the planet from a northern perspective. We're very excited about this prospect.'

And Will Whitehorn was also exploring the prospect of two RAF bases in the UK, which might be used for some summer flights – at RAF Lossiemouth in Scotland and RAF Devon. Both have restricted airspace and are outside the major commercial airline flight paths.

'Certainly, New Mexico will be the base and our headquarters – also where the spaceships will be maintained, repairs undertaken and where the bulk of the flights will take place. But I see no reason why they can't go on summer tours to suitable sites where there is a long enough runway – and the possibility of seeing the curvature of the Earth from a different angle. That could be another selling point for Virgin Galactic.'

The New Mexican wave is only just beginning.

16. BURT AND WILL GO TO WASHINGTON

A brand-new industry needed some teeth. It needed proper regulation and official approval. And it required the full blessing of America's convoluted democratic process. On 20 April 2005, a new era in space was beginning in Washington DC, and Room 2318 of Raeburn House – a warren of committee rooms and politicians' offices located on Independence Avenue, a few blocks from the Smithsonian's National Air and Space Museum, and two minutes' walk from the US Capitol – was buzzing with anticipation.

Ken Calvert, the Republican chairman from California, banged his gavel on the block promptly at 10 a.m. to open a session that would change the face of space in the United States. The House of Representatives Space and Aeronautics Subcommittee was about to hear something radical, which would take some time to sink in. But, more than that, here was a Brit coming to tell the Americans how it would work.

Calvert, a restaurant owner and property tycoon who represented the 44th Congressional district of southern California was better known by many for being caught with his pants down in a car with a hooker. But, personal foibles aside, he has been a diligent politician eager to embrace new thinking. Calvert cleared his throat. 'Good day, ladies and gentlemen. In today's hearing, we are going to examine the future of the commercial space market. We are going to have two panels. The first will examine the success of the world's first private effort to launch a person into space and to launch the hopes of our nascent commercial space industry that may lead to a robust market for space tourism.'

Sitting in front of Calvert and his officials was the first panel of the day, with Burt Rutan next to Will Whitehorn. Calvert made a special mention of the South African entrepreneur, Elon Musk, the chief executive officer of SpaceX, who was on the second panel of the day. 'I was most impressed with the work that his folks were doing when I toured his facility recently in El Segundo, California. His company is developing a new family of launch vehicles – the Falcon.'

Also appearing was John Vinter, the chairman of International Space Brokers, there to offer guidelines of what the insurance community will require for those start-up companies, Wolfgang Demisch, an aerospace financial analyst there to describe how the angel investors would look at the future risk and reward of space flight, and Dr Molly Macauley, examining what the government should do to encourage this start-up industry.

'The history of success in the commercial space arena has been spotty at best. Today, I want to see how the government can be an enabler, rather than a hindrance, to this important, hi-tech industry,' explained Calvert.

But commercial space was not entirely new, Calvert pointed out, and he went on to summarise the current situation. Satellites in orbit above the Earth were now making our modern world a much easier place to live in, to the point where they have become indispensable. Satellite dishes on tens of millions of homes beamed multi-channel sport, news and entertainment across oceans in a blink and there were over thirty commercial satellite companies around the world.

Weather satellites were now so sophisticated they could predict local micro-weather fronts, helping farmers to cut their crops more successfully. And global positioning and pinpoint tracking was helping the movement of goods and products around the world with on-time delivery.

Around the world nearly a billion people were using mobile phones, cellphones, pagers, BlackBerrys and high-speed connection to the Internet, all becoming as essential as a morning cup of tea or coffee. Most of these services and devices were using satellite relays in addition to terrestrial network technologies. Backpackers in the hills, sailors on yachts, drivers in cars, now carry lightweight, increasingly low-cost, and highly capable GPS receivers so they know exactly where they are. Satellite radio receivers are in cars, homes and boats, and hand-held satellite radios are enjoyed by the jogger slogging around the park.

The House Committee has had a chequered history of involvement – some would say meddling – in the commercial space industry, and Calvert said he was hoping to glean information that would be valuable for a NASA authorisation programme.

Burt Rutan was called to make a statement. 'Thank you for the invitation to address this important hearing. I will attempt to specifically address the subjects outlined in the invitation.'

He opened by saying the markets for a future personal space flight would take two basic forms. The first was commercial companies developing low-cost versions of the classic government booster and spacecraft concepts and then conducting commercial flights funded by passenger ticket sales.

'This activity might properly be compared to the trekking outfits that take courageous adventurers to the top of Mount Everest; the activity survives even though more than 9 per cent of those who have reached the summit have died on the mountain, with the recent rate still at 4 per cent,' pointed out Rutan.

'The safety record for all of government manned space flight is hardly better; 4 per cent fatality for those who have flown above the atmosphere, with the fatality rate for the last twenty years being much worse than the first twenty years.'

Rutan's first scenario suggested a very limited market with a sky-high ticket tag for those willing to pay millions of dollars. He believed such space systems might begin commercial flights in 4 to 6 years, flying perhaps 50 to 100 astronauts the first year with the rate topping out at maybe 300 to 500 per year.

'The second is a scenario in which the players do not find the dangers of space flight acceptable and recognise that extensive improvements in safety are more important than extensive improvements in affordability. Those that attack the problem from this viewpoint will be faced with a much greater technical challenge: the need for new innovations and breakthroughs,' he said.

He was now warming to his theme and raised his hands to explain.

'If successful, however, they will enjoy an enormous market, not one that is limited to servicing only a few courageous adventurers. It is likely that systems that come from this approach will be more like airplanes and will operate more like airplanes than the historic systems used for government manned space flight.'

He admitted that Scaled Composites' future plans could not be revealed, since they were only at a preliminary stage of technical development. But, he added, 'I can assure you that they do not involve a "scenario one" approach, since we believe a proper goal for safety is the record that was achieved during the first five years of commercial scheduled airline service which, while exposing the passengers to high risks by today's standards, was more than a hundred times as safe as government manned space flight.'

Burt then said that meeting these targets required new generic concepts; ones that would come from true research, not merely retreads of NASA's exploration plans.

'I can tell you that we do not yet have the breakthroughs that can promise adequate safety and cost for manned orbital flights. That is why our early focus will be on the suborbital personal space-flight industry. Our recent SpaceShipOne research programme did focus on the needs

for safety breakthroughs by providing an air-launched operation in which the rocket propulsion is not safety critical, and
the 'care-free re-entry' concept assures that flight control is not
safety critical for atmospheric entry,' he told the committee.

He pointed out that the airline experience had shown that
it is not just technology that provides safety, but the maturity
that comes from a high level of flight activity. Airline safety
increased by a factor of six within the first five years without
an accompanying technology increase.

'I am not able to reveal the schedule for the introduction of
our commercial systems. However, I believe that once the
revenue business begins it will likely fly as many as five
hundred astronauts the first year. By the fifth year the rate will
increase to about three thousand astronauts per year and by
the twelfth year of operations fifty to a hundred thousand
astronauts will have enjoyed that black-sky view,' he said.

Calvert looked up from his writing pad and raised his
eyebrows at this figure.

'Now that it has been shown that a small private company
can indeed conduct robust, suborbital manned flights with an
acceptable recurring cost, I do not believe that this industry
will again be hampered by the inability to raise capital. The
size of the potential market supports significant investment,'
he said.

Rutan admitted that insurance is a significant obstacle. He
agreed it would be expensive until it is shown that aggressive
safety goals are indeed being achieved. 'With maturity I expect
that safety will continuously improve, as it did with airliners,'
he added. 'Over the last thirty-three years my companies have
developed thirty-nine different manned aircraft types. All were
developed via research flight tests flown over our California
desert area and all flights were regulated by the Federal
Aviation Administration-AVR (the airplane folk, now Aviation Safety or AVS). We have never injured a test pilot, nor
put the non-involved public or their property at risk,' he said.

In spite of Rutan's exemplary record, the FAA insisted that
the Office of Commercial Space Transportation impose their
commercial launch-licence process on the last 5 flights of his

88-flight research test programme. Rutan opened up with both barrels to give the airline regulators a double blast.

'That would have been fine, except that their process bore no relation to that historically used for research testing. The AST process, focusing only on the non-involved public, just about ruined my programme.'

The FAA's Commercial Space Transportation authority (AST) was established in 1984 as the Office of Commercial Space Transportation but was transferred to the FAA in November 1995. The AST has issued over a hundred launch licences for commercial launches of orbital rockets and suborbital rockets since then. But Burt complained that it resulted in cost overruns, increasing the risk for his test pilots. He said it did not reduce the risk to the non-involved public and destroyed his 'always question, never defend' safety policy.

He pointed out that the process deals primarily with the consequence of failure, where the aircraft regulatory process deals with reducing the probability of failure. 'The regulatory process was grossly misapplied for our research tests and, worse yet, is likely to be misapplied for the regulation of the future commercial spaceliners. The most dangerous misapplication might be stifling innovation by imposing standards and design guidelines, rather than the aircraft certification process that involves testing to show safety margins.'

He told of the considerable resources he had spent developing recommendations for specific regulatory processes that could be applied to the new industry, but hadn't found any interest within the FAA. After making this point, Rutan sat down.

Will Whitehorn lodged the next statement, a bold endorsement of commercial space. But he had one overriding message he wanted to get through: safety was first, second and last.

'Chairman Calvert, Ranking Member Udall, and other members of this distinguished subcommittee, on behalf of Virgin Galactic, thank you for the opportunity to testify today. Virgin Galactic appreciates the chance to explain how, with an unwavering commitment to safety, we plan to make available and affordable an adventure of a lifetime. We are proud to be on the leading edge of the commercial space industry and honoured to have Burt Rutan as our future partner.'

He introduced himself and then handed out a bit of praise. 'At the outset, I wish to acknowledge the invaluable leadership the House Science Committee and this subcommittee provided last year for the nascent commercial space industry.

'You ensured Congress struck a proper balance in the Commercial Space Launch Amendments Act of 2004. Had it not been for that sensitivity in crafting a proper regulatory oversight regime consistent with the goal of permitting our emerging industry to realise its full potential, it is unlikely the Virgin Group would have made our considerable commitment to Virgin Galactic,' he said.

He told the committee that Virgin Galactic was a private-sector venture and received no state aid. 'Frankly, we think that is the way it should be. Entrepreneurs like Sir Richard Branson who are willing to shoulder the economic risk and challenge of commercialising space will be the most successful innovators who lead this industry and chart its course,' he went on. Whitehorn gave his view that government's proper role is regulatory, creating a climate in which entrepreneurs can translate their vision into reality, and allowing innovation to flourish.

He said safety would always remain the first priority. 'Our commitment to safety extends beyond the Virgin name, one of the best known and most valuable brands in the world. Sir Richard Branson has said that he, along with his parents, son and daughter, plan to travel in Virgin Galactic's first space flight.

'If the Federal Aviation Administration permits me to do so, I hope to be on an earlier test flight. Our commitment to safety is very real and personal to us.'

Then he stated something that has gone down in space-industry folklore: 'Safety is and will continue to be Virgin Galactic's North Star.' It was a tidy maxim – and Whitehorn truly meant it.

The subcommittee had requested information about the timetable for taking possession of the Virgin Galactic spacecraft, its first flight and expected profitability. At this time, Virgin Galactic only had a memorandum of understanding with Burt Rutan's company, Scaled Composites, to customise the SpaceShipOne vehicle for commercial use. However, Virgin had not yet formally ordered the spacecraft.

Whitehorn said US government technology-transfer issues needed to be clarified before he would place a firm order for the spacecraft. At this point, due to uncertainty about possible licensing requirements, he was still not able even to look at Scaled Composites' designs for the commercial space vehicle.

'Mr Chairman, we are not concerned about this lack of clarity on the technology licensing issue and the nominal delay it has caused to date. Like any nascent industry overseen by government oversight agencies faced with issues of first impression, we understand instances such as this are to be expected. We are continuing a robust and cordial dialogue with the US Department of Defense and other agencies that provide input on technology licensing issues. We hope a consensus can soon be reached that will clear the way for us to move forward with a formal order for Mr Rutan's spacecraft.'

Whitehorn then explained the difference between buying a commercial spaceship fleet and Virgin Atlantic's experience of buying planes from Boeing. Virgin Atlantic is a customer of both Boeing and Airbus aircraft, which is a passive process. While a buyer can request some custom features, the aircraft as designed by the manufacturer, plus the engines, are complete units and customer suggestions and requests tend to relate to the margin. Virgin Galactic's relationship with Scaled Composites was entirely different. This was an active partnership.

It would mean working closely together in designing the aircraft and sharing complementary expertise. The Virgin Galactic president told the committee it would be a symbiotic relationship where ideas and intellectual capital would be shared by the customer and manufacturer to ensure a successful product. 'This active partnership dynamic is precisely why we are so pleased to have Burt Rutan as our future partner.'

The British business was also focused on complying with the letter and spirit of the Commercial Space Launch Amendments Act of 2004. Scaled Composites would have sole responsibility to certify the spacecraft.

'However, together, we are engaged in an active dialogue with the Federal Aviation Administration on other aspects of our business. At the same time, we are designing a programme to prepare our astronauts for an incredible sensory experience and to allow them to gain the maximum from their journey to space. That programme will include training in all areas from physiological to psychological. We want to ensure our passengers have the optimum sensory experience but, even more importantly, that the operation will be undertaken with the utmost safety, consistent with safety being our absolute priority.'

Calvert asked what, if anything, the US government should be doing to encourage commercial space. Whitehorn stuck to his theme, saying that as a private business there could be mutually beneficial partnerships between NASA and private companies, especially if the US supported more public–private partnerships.

'NASA should seek opportunities to contract with private-sector manufacturers for cutting-edge designs and outside-the-box thinking. I am encouraged by signs of progress in NASA's willingness to engage with the private sector in idea sharing. This spirit of co-operation should be encouraged and broadened whenever practical to do so,' said Whitehorn.

He assured the committee that Virgin Galactic would welcome the opportunity to provide assistance to NASA on aspects of astronaut training. This could well become a lucrative area for Virgin to explore once it began flights.

Molly Macauley, an experienced expert and commentator on space and economic matters, then made some valid comments about the role of government in space commerce. There were fundamental issues about space, defence and national security that still had to be hammered out. She noted that America's commercial space policy so far had been supportive of US industry and set a good precedent for the future. Then she recommended the usefulness of demonstration or pathfinder, experimental approaches to policy.

She said that commercial space could not be a stand-alone industry and would continue to have relationships with defence and military work. Space-related markets were now more competitive than in past decades as other nations now recognised the potential. Space-transportation markets now included suppliers in Europe, China, Russia, Ukraine, Japan and India – all offering commercial launch services. Israel and Brazil also have their own launch capability.

According to the Office of Commercial Space Transportation in the Federal Aviation Administration, the US share of the worldwide commercial launch market had averaged about 35 per cent of total launches and about a third of total revenue (of a $1 bn total market in 2004, the US share was about $375 m). The total number of launches in the previous five years had been smaller than in previous years, largely due to longer-life satellites and a decline in the number of small satellites launched into non-geostationary orbit. In 2004, American companies launched 6 out of a total of 15 worldwide commercial launches – and up to 2013 the average is expected to be about 23 commercial launches per year.

Yet, as Rutan had pointed out, the arrival of a player such as Virgin Galactic would break through this figure within months.

There has been a welter of regulation from Washington in the last two decades: from the Land Remote Sensing Commercialisation Act of 1984, the Commercial Space Launch Act of 1984, the Commercial Space Launch Act Amendments of 1988, the Launch Services Purchases Act of 1990, the Commercial Space Act of 1998 – requiring the NASA Administrator to study the feasibility of privatising the space shuttle – through to the recent and most pertinent for Virgin Galactic, the newly passed Commercial Space Launch Amendments Act of 2004. The legislation, signed into law by

President Bush just before Christmas, gave the industry most of what it wanted: removal of the regulatory uncertainty surrounding suborbital spaceflight by clearly assigning jurisdiction over such vehicles to the Federal Aviation Authority's Office of Commercial Space Transportation (known by the acronym AST). The law limited the authority AST would have over the industry, allowing companies and markets to mature before the government could enact detailed regulations governing them. It allowed the licensing of private spacecraft on experimental bases and the establishment of liability in the case of an accident; and provided a legal basis for allowing private and commercial passengers to undertake space travel, establishing the concept of informed risk for space passengers.

'The objective of policy options such as these is to encourage flexibility, discourage government intervention when private institutions could suffice, and ensure a "fair playing field" between government-space and commercial-space activities,' said Macauley. 'I know from Chairman Calvert's recent comments at the twenty-first National Space Symposium this month that there is concern about sectors of the US space programme working in isolation from the others ... Our space and space-related agencies now range from the national security complex to NASA, the Department of Interior and the US Geologic Service, the Department of Commerce and the National Oceanic and Atmospheric Administration, the Federal Aviation Administration, and the Federal Communications Commission. The Departments of State and Energy, together with the Department of Commerce, are key champions of the GEOSS programme. The Department of Energy also plays a role in space power systems. To some extent, our space sectors have mutually benefited from this mix. For instance, GPS is owned and operated on the defence side but routinely used by the civil and commercial sectors. Remote sensing and earth observation information [now used by Google Earth] was championed by NASA and the infrastructure, data, R&D, data validation and information products from NASA's earth science activities over four decades are routinely used by the defence and commercial sectors. Commercial satellite telecommunications were advanced markedly by industry but are routinely used by the defence and civil sectors.'

Some steps could be taken to integrate the large scale and set out government space and space-related activity. For instance, establishing prizes for innovation for all three space sectors – civil, commercial, and national security – made sense provided all three sectors have a few desirable innovations in common.

After a recess, Elon Musk, the chief executive officer of SpaceX, was asked to give his testimony. He said SpaceX was dedicated to improving the reliability and cost of access to space for the greater purpose of helping become 'a true space-faring civilisation' – a cribbed line from John F Kennedy's famous space speech in 1962.

'All in all, I see an increasingly positive future for commercial space activities over the next five to ten years . . . The most important thing that the government should do is adopt a nurturing and supportive attitude towards new entrepreneurial efforts. In particular, the government should seek to purchase early launches as well as offer prizes for concrete achievements,' said Musk.

He singled out the success of the X Prize. 'Regarding purchasing early launches, the Defense Department has been very supportive and has done the right thing at every level, purchasing two of the four launches we have sold to date. Regrettably, however, NASA has not yet procured a launch and has provided less financial support than the Malaysian Space Agency, which has bought and paid for a flight on Falcon 1.

'As far as what the government should not do, I think it is important to minimise the regulatory burden required for space-launch activities. We should do no more than is necessary to protect the uninvolved public. It sometimes seems to me that our society is paving the road to hell one regulation at a time,' he concluded.

John Vinter, the chairman of International Space Brokers, a global insurance broking business, represents an array of satellite users, including Intelsat, Worldspace, AT&T, Bigelow, SpaceX and SES Astra in Luxembourg, Telesat Canada, Optus in Australia, Star One in Brazil, Singapore Telecom in Singapore, and others. His company also managed the space shuttle's third-party liability insurance programme for NASA. He is also the chairman of COMSTAC, the Department of Transportation's Commercial Space Transportation Advisory Committee, advising the FAA's commercial space transportation office.

'With respect to commercial space activities, we include any space activity which does not directly involve the US government as an insured. We address satellite insurance and risk-management needs from cradle to grave.

'For us, commercial space begins with the arrival of people or equipment at the various launch sites, continues through launch, deployment, testing, and on-orbit operations of satellites through the end of their expected lives. These are the areas of risk and insurance where we spend the majority of our time and where satellite owners spend the majority of their insurance money,' said Vintner.

'The launch itself is generally the riskiest and most expensive phase of any commercial space endeavour to insure. In simple terms, our objective is to cover risks of loss or damage to the satellites, including failure of the launchers, or failure of the satellite to work according to its specifications. We also provide liability coverage for damages to third parties caused by launch-related and satellite-operational accidents. Again, commercial space insurance coverage begins with the start of launch-site activities and continues through on-orbit operations. As with the satellite coverage above, activities prior to arrival at the launch site are best covered in non-space insurance markets. We also insure persons, for example, the lives of various astronauts and tourists or visitors to the space station, including individuals who fly or have flown on the shuttle.'

Vintner was asked what, as an insurance broker, he saw as the outlook for commercial space activities over the next five years.

'As brokers, we see space activities evolving and growing, albeit not very fast. The world satellite manufacturers and launch-vehicle providers have considerable excess capacity at the moment. There does not seem to be sufficient demand to absorb this excess in the near future. For the next several years, it would appear there will be approximately fifteen to twenty commercial launches per year.

'We see, however, more human activities in space, the X Prize being the first of what is expected to be a significant increase in the number of humans going into space. I have no doubt that this prize, and other incentive programmes, will generate an increase in activities, although it is hard to determine how long this will take.'

Vintner's view was that going into space was expensive and involved significant risk. He said the implications of the low-earth-orbit projects in the late 1990s adversely affected the financial markets. 'I have no doubt that the financial community will demand sound business plans before advancing significant sums of money,' he said. This, after all, was an insurer talking.

The arguments were obviously being won, though.

Meantime, Whitehorn and Attenborough continued their punishing schedule to win the public-relations battle in America. On 21 May 2005, Virgin Galactic's president was able to reveal a few more nuggets. During a session of the National Space Society's International Space Development Conference in Arlington, Virginia, he offered a taster. 'What we're doing is giving you a two-hour experience of going into space that will only require a few days of training.'

He explained that the experience would begin immediately after customers arrived at Virgin's training facility, initially to be located in Mojave, California. The cabins of SpaceShipTwo and its carrier aircraft – White Knight Two, or 'Eve', named after Richard Branson's mother – will be identical, and the new group of customers will be given the opportunity to fly on the aircraft during the launch of the previous group of tourists.

'They'll get the excitement of going up to 50,000 feet, which not many of us have been able to do,' said Attenborough, 'and they'll also be able to watch from prime position the launch of the spaceship that's going before them.' This would be a spectacular vantage point.

Other aspects of the training will include extensive time in simulators, so that 'everything will feel familiar' when it comes time to take the actual flight. There would also be flights in fast, light aircraft to acclimatise people to the G-forces they would feel during the actual flight. Customers will also be trained to operate the 'personal communications console' they will use during the flight to record their experience.

The Virgin team were keen to discourage personal cameras and camcorders that might float around in the cabin – but how can you tell someone who has forked out $200,000 not to bring their pocket digital camera – or mobile phone?

SpaceShipTwo will fly at a peak altitude of 360,000–400,000 ft and this trajectory will give passengers about four to five minutes of weightlessness. At this stage, Virgin were considering using a tether that would make it easy for passengers to return to their seats when the G-forces of re-entry build. But that was later thrown out.

Whitehorn told the conference that Virgin Galactic estimated it would take about 450 people into space during its first year of operations, roughly the same number that have flown in space since 1961. 'We will double that number in year two, flying over a thousand people,' he added.

Then he repeated the commitment he had given the science subcommittee a few weeks earlier: 'The North Star of this project is safety. Safety is really at the top of people's lists as to why they think they're interested in flying a suborbital space flight.'

It was that commitment to safety, Whitehorn explained, that led Virgin to back an air-launched system instead of a ground-launched system. 'If you're going to commercialise this business, you've got to be able to take thousands of people into space safely,' he said. 'Any system which is ground-based has intrinsic issues with safety which an air-launched system does not have.'

Whitehorn's acutest fear was if a 'cowboy' venture managed to start launching paying passengers into space – then had a serious accident. 'One of the biggest risks we face is if someone in the next three years decides to put somebody into space using ground-based rocketry and they have an accident,' he said. 'Because the most likely thing that would result in is AST being forced by Congress to shut down the whole programme. If that happens, I think we have a real problem on our hands.'

This was an issue beyond the power of Whitehorn and Rutan. They could only get on with their own plans and, instead of worrying about regulatory issues, focus on developing their safe vehicles and business plans. While the industry had won the first major battle, it still faced a mountain of work over crew and passenger safety, both in their dealings with AST as well as among themselves. This ding-dong battle for regulation would continue until the first flights were on the launch runway.

17. THE GLOBALFLYER

New technology and lightweight materials are an essential part of Virgin Galactic's plan to take passengers into space. The GlobalFlyer single-pilot nonstop round-the-world-plus challenge was a significant feather in the cap, and remains an important test-bed towards space flight. Here was another proving ground for the composite technology being pioneered by Burt Rutan, technology that would be the backbone of taking people up in SpaceShipTwo.

Burt Rutan, speaking in 2006 at the twentieth anniversary celebration of the Voyager round-the-world flight at Mojave airport, said: 'We concluded back in the early 1980s two significant things. One, that we couldn't afford to buy a jet engine, and two, that it was impossible to fly around the world with a jet engine, so I think the technology improvement that's happened in these twenty years have been great. Because the airplanes were designed twenty years apart, it is clear that we did increase the structural efficiency. The Virgin Atlantic GlobalFlyer structure weighed 7 per cent of the gross weight while the Voyager was 9 per cent. And we also improved the aerodynamic efficiency – we had to do that because the jet engine was not anywhere near as efficient as the Voyager engine. And if you were sitting for sixty-seven hours above the weather on autopilot, the book wouldn't be nearly as exciting.'

But Burt was wrong about the action. Steve Fossett's flight was packed with high drama and certainly wasn't a stroll in the park. Steve's route was to take him from Kennedy Space Center in Florida across the North Atlantic. He would then fly over North Africa and on to the Middle East, then cross over India, China, Japan and the Pacific. He would aim for Mexico and back across the US for his final leg, which would require him to cross the Atlantic for a second time. In order to secure the record he had to fly over Shannon in Ireland and onto his final destination, Kent International Airport in England.

In announcing his intentions, Fossett said: 'As many of you know, I've attempted (and been lucky enough to succeed) in a few world record attempts in my time, but to achieve the ultimate distance ever flown would be a dream come true.'

In March 2005, the Virgin Atlantic GlobalFlyer lost a lot of fuel during her round-the-world flight but still broke the absolute nonstop world record. The fuel fraction of Global-Flyer was fundamental to the success. On takeoff 83 per cent of the total weight was kerosene – with the remaining 17 per cent being the aircraft and the pilot. This was an unprecedented figure in aviation. The fuel is distributed in thirteen tanks, which change the aerodynamics of the plane the longer it flies and as the weight lessens.

By ironing out fuel-tank problems and making the plane even more efficient, Fossett wanted to go even further. 'I believe that these problems have been rectified, which is why I am determined to test this theory by attempting the longest flight ever flown by a balloon, airship or airplane. This might seem a bit crazy, but I've been called crazy before! Everyone who has attempted to set or break world records will know that if it isn't a true test and challenge there's no point in taking it on,' said Fossett.

'I can only hope that the speed of the jet streams performs better than expected in certain parts of the route and those predicted strong remain so – it would be great to land in the homeland of Phileas Fogg in the predicted eighty hours having, in Virgin Atlantic GlobalFlyer, just achieved the ultimate distance ever,' he added.

Fossett's flight in March 2005 achieved the first ever solo, nonstop circumnavigation of the world, despite losing over 3000 lb of fuel due to a design problem with the fuel vents. On Thursday, 3 March 2005, at 1.37 p.m. local time, a tired and bleary-eyed Fossett touched down at Salina Municipal Airport in Kansas. He had taken 67 hours, 1 minute and 10 seconds to fly 22,396 miles (36,912.68 km) around the globe at an average speed of 342.2 mph. He was greeted by a boisterous crowd of aviation enthusiasts, among them his wife Peggy and Sir Richard Branson, who doused him in a magnum of champagne. Amid the plaudits and emails was a letter from Dick Rutan which said: 'My hat's off to Steve Fossett for a picture-perfect landing in Kansas after his gruelling and record-setting nonstop world flight. I've said many times I hoped Voyager's record would be broken in my lifetime, and I congratulate Steve ... I also need to congratulate my brother, Burt Rutan, for yet another incredible, beautiful and uniquely successful design.'

Real-time flight data and reports had been fired back to mission control, based in Virgin Atlantic Airways' head-quarters in Crawley, for analysis throughout the flight to assess the problem and whether it could be rectified. Director Kevin Stass was in constant communication with pilot Steve Fossett throughout the flight and it went well. It was a superb achievement, but Fossett – with more than fifty airplane, ballooning, gliding and sailing records – wanted one more.

On 19 December 2005, Fossett announced that NASA's Kennedy Space Center would be the launch site for the Ultimate Flight, which would see Fossett pilot the GlobalFlyer aircraft to set the record for the longest flight of all time. The existing record for a plane was held by the Voyager aircraft, which flew for 24,987 miles (40,212 km) in 1986. Burt Rutan had designed and built that propellor-driven aircraft, which his brother, Dick Rutan, and Jeana Yeager flew on a nine-day around-the-world flight in 1986. The longest flight by any aircraft was the Breitling Orbiter balloon, which flew for 25,361 miles (40,814 km) in 1999.

Now Fossett's aim was to get past 26,000 miles, but he would beat the record if he managed to travel 25,977 miles (41,806 km). For a 59-year-old man, however fit he was, this was an incredible challenge pushing him to very edge of human endurance. To make the ultimate record, Fossett needed to fly the distance from the start, through six intermediate waypoints to a declared finishing point. These waypoints must be at least an average of 5,000 km apart and the route is set by air-traffic-control standards. This meant a lot of wasted miles that would not count for the record. So Kevin Stass and his team at Kansas State University managed to work out a more direct route crossing the waypoints to make the world record distance.

The mission was set to go on 7 February 2006, but Fossett was forced to postpone the flight for 24 hours when a fuel leak was discovered just half an hour before the planned takeoff. But the mechanical problem was quickly patched up and tested by the engineering team.

'This is obviously very disappointing news for the whole project team,' said Fossett. 'As anyone knows who undertakes challenges of this nature, there are always obstacles to overcome, whether it is the weather or mechanical issues. I have total confidence in the team to rectify this problem so we can launch the record attempt as soon as the fix has been undertaken and the weather permits. Mission Control is continuing to analyse both the ground forecasts for takeoff and the jet streams around the world in order to identify the next possible takeoff opportunity.'

Jon Karkow, the chief engineer at Scaled Composites, admitted, 'This is a disappointing but minor issue that luckily can be easily rectified within one day. To put it simply we have experienced a fuel leak in the plumbing. The joint between the fuel vent line and its tank has leaked. This is a bonded joint, which uses a sealer to keep it from leaking; unfortunately on this occasion the seal failed. The leaking joint on the left aft boom tank is part of the newly installed boom vent system, a system designed to prevent similar fuel loss which occurred in the Virgin Atlantic GlobalFlyer's successful round-the-world record attempt last year. As a result this particular joint has never seen fuel before, as this morning was the first time the aircraft had been fully fuelled. It is a simple process of replacing the failed seal and we are confident that we will have no further problems.'

Richard Branson was still smiling and relaxed. 'This is an experimental aircraft and it is just one of those things. We will fix this particular problem in twenty-four hours and then we will have to work on fixing the weather.'

Next day, 8 February 2006, Virgin Atlantic GlobalFlyer mission control announced that Fossett had successfully taken off from the Kennedy Space Center in Florida at 7.22 a.m. Eastern Standard Time.

The ground conditions were perfect for takeoff with north-to-south head winds and a temperature of 47°F/8°C. But Fossett strained to get the plane – groaning with fuel in every part of its structure – off the ground. The engine hit its maximum temperature of nearly 1,000°C as it lifted off with only a little runway to spare. And as it soared it hit a flock of sea birds – thankfully the plane withstood the impact.

After a fairly uneventful first 50 hours and 48 minutes, Fossett had covered 19,019 miles at a sedate 242 mph. The GlobalFlyer had picked up great wind speeds across the Pacific. He had 4,000 lb of fuel left as he crossed the California coast. After sixteen hours of sea and sky, Steve was glad to see terra firma, and reported: 'It will be such a relief to see land again – the Pacific sure is a big ocean! It will be great to see America again after over fifty hours of flight. The Pacific gave me time to liaise fully with mission control, after the strenuous time I had with turbulence over India.'

He passed over southern California, on to San Diego, Phoenix, Midland Texas, Louisiana, St Petersburg and Florida. Winds over both the United States and the Atlantic remained very weak. 'I know the guys at mission control are looking at every available jet stream to pick the best speeds across the Atlantic as possible. In the meantime, I'm going to attempt to sit back and enjoy the view!' he said. It was the lull before the storm.

When the Virgin Atlantic GlobalFlyer headed over Big Spring, Texas, Steve Fossett had flown 22,383 miles – already a fantastic feat. The public interest in his flight was phenomenal. At its peak, there were five million hits an hour on the website, and an email was received every twelve seconds. The public were willing him to succeed. The Atlantic Ocean was the last big stretch – and the plane was straining as it became lighter and lighter as the fuel burned away. Eventually, Ireland was in sight and then it was the home straight into the United Kingdom.

The mission control guys had found a good flight path over the Atlantic, but deep fatigue was setting in. And Fossett would need all his wits and years of flying skill to combat an alarming situation. As he prepared for his 40,000 ft descent into Kent International Airport at Manston, the GlobalFlyer experienced a large-scale electrical malfunction. All the systems went down – virtually turning the lightweight composite plane into a glider.

Steve grabbed his radio to make a Mayday call. Jon Karkow and mission control became deeply anxious and Steve's instinct was that he would not be able to make it all the way to Kent. He decided instead to seek alternative landing options in either Cardiff or Bournemouth. Having landed at Bournemouth Airport in the past and knowing it was downwind, Steve opted to attempt his landing there. However, it was not going to be easy. He couldn't see out of the cockpit window because of thick ice, but with only 200 lb of fuel left and with an aircraft in major electrical malfunction, Steve was unable to circle around and wait for the ice to disappear. He had to glide in immediately, even if this meant bringing in the plane almost blind. As the plane hit the ground, two tyres burst, making an already hazardous landing even more dangerous.

But he had broken the record. His final distance was logged at 25,766 miles (41,467 km); it had taken him 76 hours, 42 minutes and 55 seconds, which broke Dick Rutan and Jeana Yeager's record by 780 miles. Despite extreme sleep deprivation, Steve was elated – 'I'm a very happy guy,' he said – before stepping up to talk to his friends, well-wishers and the press about the drama. 'It was too exciting a finish. There were many obstacles to overcome from the moment I took off from Kennedy Space Center, from the challenging takeoff, difficult cockpit conditions for the early part of the flight, severe turbulence over India and constant concerns over the weakness of the jet streams due to the less than favourable weather patterns around the world. But never during all of this would I have believed that forty-five minutes out of Kent I would be in an emergency situation the like of which I have rarely experienced before.'

It was a gigantic achievement – despite the electrical failure, the robust composite structure passed its test with flying colours. This was great news for Scaled Composites.

18. GRAND DESIGNS ON SPACE

The original American astronauts were crammed into a tiny capsule and called it 'Spam in a Can'. While some space tourists would like the full Gemini experience – with full spacesuits – the idea for Virgin Galactic was to make the trip with more panache and style.

And many of the world's leading designers were soon clamouring to have their imprint in space. But it was the French legend Philippe Starck who wanted it most. Starck was having dinner with Richard Branson and rock star Peter Gabriel, one of the original members of Genesis, back in Christmas 2004.

Over brandy, Richard made an off-the-cuff offer: 'Come and work with us at Virgin Galactic.'

Starck smiled. He didn't need a lot of convincing.

'OK. I'll give it a try.'

There was a nervous laugh from one other person in the room. Will Whitehorn, the chairman of Virgin Galactic, had a problem. Money, or lack of it. He was on a tight budget.

'But, Philippe, we couldn't really afford your fees,' White-horn revealed to the Frenchman.

'Then I'll do it for nothing.'

Not quite for nothing. A few days later Starck went to see Stephen Attenborough, who was still working in Virgin Management's office before the move to Half Moon Street. Attenborough was delighted with Starck's involvement. 'I asked him what he wanted to get out of it and he came up with an extraordinary offer. He said that this project was part of man's evolution – and one of the most important things that was happening.'

A mutual arrangement – involving a space flight – was made very quickly, recalls Starck. 'Soon after meeting Richard and Stephen, I started as art director and became the angel of the spirit of the project,' he says.

Starck buried himself in his studio for a few months, revisiting some themes and thinking about concepts. He had become consumed by the challenge. While his work is normally associated with spatial form, he has been obsessed by flying and outer space since he was a child growing up in Paris. As his work grew in stature, he returned to spacelike themes. Back in 1982, his sketches included satellites and the Earth floating in space. Many of his large exhibits featured the globe. In 1990, his *Moondog* project in Tokyo was a stairway to the heavens in black and white chess squares with the green tip of a spaceship pointing out the top at the stars.

Across Europe, Starck is renowned for the furniture that bears his name – especially his chairs. He is a commercial designer who has invented thousands of product designs. Yet his architectural work – although he is not a trained architect – is also world class. Indeed his celebrity in Japan is such that he is stopped in the street there. His design has been embedded in so many postmodern designs from the toothbrush to Alessi kitchenware through to Olympic Flame torches and bathroom taps. He worked for the French electrical company Thomson, where he invented the slogan: 'From technology, with love'. His original work is ubiquitous in our modern everyday world – and he has many imitators.

On science, he says: 'I didn't study science but it's the only thing that interests me. Through theoretical rudiments, I seek the first flickering of comprehension, moving towards an understanding of everything that surrounds us. There lies the nobility of thought . . . I understand science as an intuition that can, perhaps, be verified.'

But he recalls the comments of one of the fathers of modern design, Raymond Loewy, who said in the 1950s, 'Ugliness doesn't sell well.' Starck takes this to the next level and says a designer's duty is to question whether a product has the right to exist in the first place.

'One of the most positive things a designer can do is refuse to do anything. This isn't always easy. He should refuse when the object already exists and functions perfectly well.'

So Starck wouldn't *refuse* the call from Branson. Indeed, Branson was a serious admirer of Starck's work. Virgin Holidays singled out the Hudson Hotel in New York, designed in 2000 by Starck, as one of its preferred travel destinations. From its Le Corbusier functional escalator entrance, to its dark mahogany reception areas with its canopy of ivy, through to the spacious ersatz library with its deep leather seats, black-and-white photographs of cows wearing hats and a mock gantry of leather-bound books, and the rococo Versace-style Hudson bar, it suited Virgin's aspirational travel market to a T. Each compact room – starting at $350 a night – is a mini-sanctuary of Starck style with its wood panelling, Archimoon lights and starched white linen bedding. Add the extra loud hip-hop music in the open-air bar and the lifts, and it became a must-visit urban stopping place for the discerningly hip consumer. The union of Branson's Virgin Galactic and Philippe Starck was a marriage made under the stars. Starck's first task was to help Virgin Galactic rebrand with something startling and original. It beat every expectation.

On 7 March 2005, Starck arranged a meeting with the Virgin Galactic executives; Will Whitehorn, Stephen Attenborough and Susan Newsam (arriving on the first day of her new job as marketing director) were whisked off across the

Thames to a meeting at Chiswick station with GBH Design – or Gregorybonnerhale – who also work for Puma and Eurostar. Mark Bonner had collaborated with Starck on a multitude of projects before, but this was literally out of this world.

Starck unveiled his thinking. Here was a strikingly beautiful logo, using a close-up photograph of Richard Branson's right iris in a blue and black image. It also looks like the aura of the world escaping into space – or a nebula. It had a multitude of explanations. The Virgin Galactic space team were bowled over. Above all though, the eye looking out represented Starck's explanation of the human necessity for space travel.

'We know that in four billion years' time the sun will implode and this world will explode. Ppoophh!' He expresses an explosion with his open hands. 'That's why I think that somewhere in our DNA we know that we are obliged to fly – to escape,' he explains in New York in September 2006.

'This project is not a joke. It is the first step for freedom for everybody to be part of the beautiful project of civilisation. A civilisation based on intelligence. This project is about the inquiry of the human mind and its natural quest to explore. It is in-built in us and we cannot deny this urge,' he adds with some sternness.

'I was born in all this. My father, Andre, had a plane factory called Starck. My father was the Burt Rutan of the last generation and my bed was under his drawing board. It was a small aircraft company like Burt Rutan's start-up. I witnessed invention and my father brought some revolution in aircraft design. So, for me, it's completely logical and coherent to be working with Virgin Galactic. I feel it in my DNA,' he asserts with conviction.

For Starck, science fiction has also had a deeper influence. From his early days he found interest in the contrast of darkness and the light in the universe. 'It's not a surprise for me, all that. It's the natural speed of things. The reality *is* the reality. It is what is out there,' he says, pointing skywards.

The Virgin Galactic Eye and the Branson iris quickly became the symbol on everything – and also the architectural shape and device for the spaceports to come. As art director, Starck enlisted GBH to carry out his thinking to the next stage for Virgin Galactic – incorporating the Eye into the spaceship itself.

'For me, I'm not interested in the rocket itself or the material of the project but by the vision. Richard had the vision and that is *fantastique*. That's why it is the eye of Richard and the spaceport will be the Eye looking up into the sky and everything about this means . . . vision. The thing we need in our society is vision because today everyone looks at their feet and doesn't believe the beautiful story.'

The Galactic space team were convinced – but Philippe Starck still had to persuade the rest of the Virgin Management team at their board meeting in Necker Island in the Caribbean. The logo was a departure from the distinctive red Virgin signature branding – and with this name so important, it would need everyone's approval.

Sir Richard recalls, 'I heard nothing for a while, and then after a couple of months, Philippe enthusiastically called me up with Burt Rutan. They met me at Necker and Philippe laid out on a billiard table the complete collection of remarkable logos, based around a nebula, combining a human eye with its pupil. The black hole represented space, which suggests "the future and beyond". There were several different images of eyes. With a smile, Philippe pointed to the blue one. "That's yours, Richard," he said.'

It was given the thumbs-up, and Sir Richard asked for the logo to be placed immediately on the sail of the new catamaran used for Virgin's well-heeled holidaymakers and on the tail of Virgin's chartered corporate jet.

Virgin Galactic required some extra stardust to keep the marketing momentum – and the press coverage – going. The businesses needed an event to reignite stage two, and an opportunity came in New York in late September 2006.

'This was one of the largest projects conceived and put on by Virgin Galactic. It involved months and months of preparation and pre-production,' says Susan Newsam. 'We knew that we needed something fresh and exciting to show our customers. A lot of them had signed up in good faith but there was nothing concrete to show them other than the original video of Brian Binnie in SpaceShipOne during the X Prize. We thought it would be wonderful to show them what the interior of the new SpaceShipTwo would look like, and actually allow them to sit in the seats.'

So Virgin Galactic set about trying to look at the experience with Starck setting the design parameter of the conceptual interior. It was to be as clean and simple as possible.

'We wanted to work closely with British product designers Seymourpowell to build a full-scale conceptual mock-up of the interior,' says Newsam, a former university contemporary of Whitehorn who was pulled in because of her widespread experience working for major European consumer brands.

Dick Powell resembles a classical English thespian, dressed in his denim two-piece suit and sporting a regal head of wavy grey hair. He was a boyhood friend of Sir Richard Branson and recalls playing a spaceship game with him when they lived nearby.

'Dick and I used to argue about *Dan Dare* comics when we were a few years old. We were best friends then,' recalls Branson.

Now Powell is a leading English product designer able to turn a few sketches on paper into something more meaningful. He faced a major headache from day one. How was he going to get six people into the cabin space? The dimensions were extremely tight yet there was a need to ensure that Virgin Galactic's paying tourists had enough room to experience the joys of weightlessness. This, after all, was why the founding Virgin astronauts were willing to fork out $200,000 for the equivalent of the five minutes of zero gravity. It was worth its weightlessness in gold.

'We couldn't really see a solution unless we had more space,' says Powell. But there was no more room. The product designers were given specific dimensions and airplane designer Burt Rutan was adamant. Few people dared argue with Burt and he was keeping his own final measurement to himself. But one thing was clear: there would be no more room.

'We were guessing a lot of the time. We were given the size and I think what we've now got is very close to what it will be like. There were only a limited number of options with the space and the number of people,' admits Powell.

Seymourpowell built an interior with six passenger seats in three rows of two. The seats were ergonomically designed to slide back from their sixty-degree upright angle for the takeoff to an almost horizontal angle as the astronauts reached space and were able to enjoy their few minutes of weightlessness.

'It was a hell of a problem. The passengers need to be upright after lift-off and then prone when coming back in to combat the G-force pressures of re-entry. The only solution we could see in the available space was our reclining seat,' says Powell.

'The front passenger seat goes under the bulkhead. It will be a comfortable but fairly tight environment. We needed to ensure that the astronauts maximise their experience of being weightless. You want to clear the space as much as you can. Everything is geared towards this, so even the windows have grips around them, so you can pull yourself about.

'There is an overlap in the seats and a space tourist's feet are tucked in behind the person in front. They will float free during weightlessness. It took a lot of working out. But this has been an inspirational project for all of us and it has been exciting working with Philippe at the briefing stage. His clarity of vision has saved us a lot of time. The Virgin Galactic people were very open to ideas in what is a paradigm-shifting spacecraft.'

Seymourpowell's team machined the original mock-up seats from foam at their Fulham workshop in Lillie Road. The Galactic team went along to see what it looked like. There was a lot of discussion about portholes in the floor before it gained the OK. The spacecraft will have fifteen triple-glassed, pressurised windows, excluding the cockpit window, nine of them seventeen inches in diameter and six at thirteen inches. Seymourpowell was also asked to create an animated video and set about capturing material for what was expected to be a mini-blockbuster.

As September's showcase event in New York approached, the designers and the mock-up builders were decidedly twitchy and frantically trying to bolt down the logistics. The animators – working on their PowerMacs – were anxious to add touches of finesse to their digital design, but time was running out. There was a lot riding on the success of this event, with a planeload of Fleet Street's finest supping Bombay Sapphire cocktails on a Virgin Atlantic Airbus 340-600 on its way to JFK airport.

But one person, known to the Virgin insiders as 'God', was pulling together all the pieces to create the 'Wow!' factor. Instead of the vast, echoey airport hangar of New York City's Javits Center, it might have been more at home a few blocks away in Broadway's theatreland.

Miles Peckham is chief executive of a niche event-management company called Watermark Event Management, based in the appropriately named Fort Wallington, Military Road, in Fareham, Hampshire. Peckham is an experienced Virgin hand, having worked with Virgin Atlantic since the eve of the classic FA Cup Final of 1990 when a panicking executive, Chris Moss, asked Peckham's company to help make a big noise with flags, balloons and colours for Crystal Palace, then being sponsored by Virgin Atlantic. It obviously worked. The Wembley match against favourites Manchester United ended a thrilling 3–3, with Palace sadly losing the replay. Since then Miles has been the man to turn to whenever Virgin wanted to launch a new route, working in Jamaica, Mumbai and Cuba.

'I've done all of the original events for Virgin Galactic, including the launch at the Science Museum in London. I was also involved in the GlobalFlyer project, which was sponsored by Virgin Atlantic, and spent some time in the Mojave Desert with Scaled Composites for the roll-out,' says Peckham.

Then when adventurer Steve Fossett went for the nonstop round-the-world record in March 2006 with Virgin Atlantic GlobalFlyer, Peckham and his Watermark team built the mission control centre in student halls of residence at Kansas State University.

'There's a lot of theatre involved. Richard likes a bit of theatre, indeed we all like a bit of theatre,' says Miles. 'I like to do a reveal. It's more fun if you can do it with something dramatic to show. Richard and Virgin like this. You need to create some magic because sometimes what you are going to show is strong but not enough of the finished article. You need to do an unveiling or a reveal or a show of some sort to bring it to life – and make it media-friendly. You won't grab worldwide television unless you put on a show. And it doesn't get much better than a spacecraft.'

Television needs moving pictures to explain stories and Peckham knows this better than anyone in the business. 'We look at what is going to be televisually friendly. And also for the snappers in the national newspapers you want to capture it in a single image too. What we are looking for is how to tell the story in one glance, so that when the curtain goes down, what do you see: you see SpaceShipOne, which is quite small, and then you see SpaceShipTwo, which is three times bigger. You see Virgin logos and you see Richard Branson. In one glance, you can tell the story: Branson, new spaceship, much bigger and bigger investment. Our brief was simply to demonstrate how the new SpaceShipTwo will be three times bigger than SpaceShipOne. It is going to be so much bigger and more exciting.'

So Miles wanted to get a replica of SpaceShipOne and hang it high up in the ceiling over the mock-up of SpaceShipTwo, so the audience could compare like for like. The team looked at having a mock-up made in England and then shipped to America, but the cost was too high and the timescale too tight. So Global Effects in Los Angeles, a Hollywood film-set design company, specialising in space sets and NASA replica suits, were commissioned to build the 25 ft mock-up plane and have it shipped across the US to New York.

'We then built a two-dimensional façade for the full length of the craft, which was sixty feet long. The façade was printed in the UK and brought over as air cargo but the large mock-up had to be brought by road. When it arrived on Monday, it was then placed under SpaceShipOne, which was dangling from the roof of the conference centre,' explains Peckham.

The mock-up – designed by Dick Powell – was to be taken by trailer from Laurel Canyon Boulevard, North Hollywood, and afterwards begin a tour of shows right across America, Europe, the Far East and Australia. 'It was a very tight deadline. By the time Seymourpowell had finalised their designs and got artwork for the exterior, it left about a month to put the whole thing together. But that's the nature of the business,' says Miles.

Increasingly the question of damage to the environment was something the incipient commercial space tourism industry had to be seen facing up to. Surely blasting rockets into space is hugely damaging, queried many people. It was a point that was exercising Virgin Galactic. For Whitehorn and his team there was no chance the public could possibly be convinced that firing old-style rockets into space was a good thing. Climate change and global warming have become mainstream issues – and a shrewd consumer brand like Virgin had to ensure it was ahead of the curve. On the Virgin Atlantic flight to America, Al Gore's film *An Inconvenient Truth* was eloquently setting out the challenges facing humankind. The former US vice president was telling us, 'Humanity is sitting on a ticking time bomb – and if the vast majority of the world's scientists are right, we have just ten years to avert a major catastrophe that could send our entire planet into a tail-spin of epic destruction.'

Indeed, Sir Richard Branson was becoming a convert. He was unable to make Al Gore's presentation at the British Film Institute's IMAX cinema in London, so the American states-man offered to do a personal presentation at Sir Richard's Holland Park home.

'It was quite an experience having a brilliant communicator like Al Gore give me a personal PowerPoint presentation. Not only was it one of the best presentations I have ever seen in my life, but it was also profoundly disturbing to become aware that we are potentially facing the end of the world as we know it,' he recalls.

The presentation was followed by intense discussion and Steve Howard, the environmental physicist who runs the international Climate Group, asked for Virgin's support and leadership. 'We need to make people confident that this is a problem that could be solved,' he said.

'We really have no choice in the matter: we *have* to do it,' replied Branson. 'The Climate Group has set out to build a powerful constituency that could take that message out there and we can help change people's perception and drive these necessary changes.'

Al Gore was delighted. He said, 'Richard, you and Virgin are icons of originality and innovation. You can help to lead the way in dealing with climate change. It has to be done from the top down, instead of from the bottom up on a grassroots level, as before.' The Virgin Group's environmental response was immediate and impressive.

Yet the frivolously wealthy were considering trips into suborbital space. This wasn't going to be a good marketing image for Virgin Galactic to resolve. Sir Richard needed to go into turbo-charged overdrive to sell his message. And he did. He also had to prove he believed it passionately too, and that would take a lot longer.

On 21 September 2006, Sir Richard, standing next to former US president Bill Clinton, pledged to commit all the profits from his transport group over the next ten years to combating global warming – a sum estimated to be $3 bn.

At the Clinton Global Initiative in New York he said: 'Our generation has inherited an incredibly beautiful world from our parents and they from their parents. We must not be the generation responsible for irreversibly damaging the environment.'

He explained that he had been persuaded by Al Gore, and that Ted Turner, the American media mogul, had persuaded him to build a refinery to make cleaner fuels. On CNBC's *Power Lunch* programme, he repeated his green pledge. 'Obviously we are in the transportation business and we do our fair share of spewing out CO_2 . . . We are pledging that any money that comes back to the group in the form of dividends, share sales or flotation, that a hundred per cent will be invested in tackling global warming. We expect over the next ten years to put aside around three billion dollars.'

He was asked by anchor Sue Herera where he would like to see the profits channelled. 'Well, the big danger . . . and the things that create CO_2, are coal, cars, lorries, and air conditioning and planes. So what we've got to do is come up with alternative fuels. Using some of this will be completely experimental and some we can get a return in by investing in new fuels. We will be [building] ethanol plants and [using] cellulosic ethanol and neutanol, which is more powerful than ethanol. And we are researching new fuels for planes. We are trying to invest in alternative fuels for the future,' he told her once his rogue earpiece was fixed.

It all made huge headlines. Then, less than a week later, on 27 September, Branson was ready to unveil another initiative to reduce carbon emissions from aviation by up to 25 per cent. He was keen to be driving the environmental debate. His four airlines use around 700 million gallons of fuel a year and he wanted to replace this with plant-based bio-ethanol over the next six years.

It was another mild autumn day in New York and the British press pack had arrived and been fed and watered for the evening. Now they headed south in a coach from the Philippe Starck-designed Hudson hotel, joined by television crews and American print journalists. Will Whitehorn, who had flown in at 8 a.m., looked rather tired and had a hacking cough, which later turned to laryngitis. He'd spent the previous day in San Francisco with one of the US's most famous venture capitalists, Vinod Khosla, the man who founded Sun Microsystems and became partner of the famed Silicon Valley venture-capitalist house, Kleiner, Perkins, Caufield and Byers. Khosla was now a committed convert to green energy and a joint investor with Virgin in their bio-ethanol plant. Will was excited about being able to enjoy some intellect sparring with the world's leading investors.

At a press conference at Soho House, on Ninth Avenue, Branson proposed starting grids for planes at airports and a method of landing planes called 'continuous descent' approach, which meant a saving in fuels. Some wags wondered if this wasn't what happened already when planes landed. The billionaire entrepreneur was hitting back, pointing the finger at Europe's air-traffic-control system, which he claimed was 'punishing the environment'. While there were 35 separate traffic-control organisations in Europe, there was a single one for the whole of the US. But, as one aviation correspondent suggested to Branson, these were all well-paid government jobs in countries across Europe and it would be difficult to break this bond without huge trade union troubles. Nevertheless, the Virgin Atlantic airline was working to pull together the airlines to make commercial flying more environmentally friendly.

It was all a little too tame for the press. Even the appearance – by massive video satellite – of Arnold Schwarzenegger, the governor of California and former actor, didn't seem to whet the appetite of the media. While the trade press asked earnest questions, the tabloid hacks from London were stumped for a strong line for their news desks. The press had a firm eye on Virgin Galactic's announcement due the following day; they knew something more sexy was still to come.

Meanwhile, down at the NextFest, an event sponsored by the Conde Nast technology magazine *Wired*, flight cases and lights were being unpacked and temporary stands arranged. The Las Vegas-based trade-show haulage company, TWI Group, had delivered the mock-up safely. Two drivers had driven twenty hours each, with a four-hour break, to deliver SpaceShipTwo from Los Angeles, leaving Laurel Canyon late on Friday night and arriving on Monday afternoon. Steve Barry, the president and CEO, who delivers full-scale helicopters and airplanes to aerospace exhibitions all over the world, wasn't going to let down his newest customer.

'I have taken charge of this personally,' he said. 'This was a 4,000-mile trip for our 51-inch RGB-EX (removable goosehead) truck, while we've also shipped in a replica of SpaceShipOne from St Louis.'

Barry landed the Virgin contract after seeing the SpaceShipTwo model on the Virgin Galactic stand at the Farnborough Air Show a few weeks earlier. He wanted one for his collection and started chatting about his family-run business – specialists in exhibitions and trade shows – and a smart-thinking Virgin executive signed him up for the job.

On the Wednesday, the journalists had the afternoon to file their stories and edit their tapes, while the last-minute work was being done and a full dress rehearsal taking place at the vast Jacob Javits Center on West 34th Street. The Virgin press team had booked a late-night transvestite karaoke bar for the journalists, the kind of crazy event that's part of the Virgin media legend.

When Sir Richard arrived at Lucky Cheng's, he met his mother and father negotiating the stairs. They had flown in along with the press party. He embraced them both, saying, 'I've just seen the mock-up. It looks great, I'm very pleased with it.'

He was then greeted like a long-lost friend by a hostess in drag. Richard Branson had obviously enjoyed the Lucky Cheng's chop suey many times before. After jugs of toxic cocktails, the karaoke was in full swing with Virgin Galactic's personnel enjoying duets with the girls. Perhaps it wasn't such a wise idea for the gregarious Whitehorn, who was sounding hoarser by the hour.

It was a fiendishly early start for everyone next morning. It was still dark, but the weather was mild. One or two journalists looked rather peaky as they stepped onto the coach at 5.45 a.m., but Virgin's bubbly head of press, Jackie McQuillan – with the slightly tongue-in-cheek title of Head of Human and Inter-Planetary Media Relations – appeared frighteningly effervescent. A serious party animal and an Irish darling through and through, Jackie is a top PR more than able to match Fleet Street's finest, margarita for margarita. Terri Razzell, now Will's personal assistant, was concerned about her boss's worsening sore throat and rushed off to find him some lozenges. Her unofficial motto for her boss was: where there's a Will, there's a PA.

The bus left the Hudson and headed to the conference centre, where a crowd was already gathering around the Virgin Galactic stand, draped in massive black curtains. About a hundred people were milling around but the rest of the vast show arena was deserted.

The Apollo 13 astronaut, a now white-haired Buzz Aldrin, was standing holding court, a small group of admirers listening to him talk about bio-fuels.

Along with all the Virgin Galactic staff, there was Dick Powell and his team, Philippe Starck, and the new travel agents who had been appointed to sell the Virgin Galactic experience to well-heeled space tourists. Travel specialists, such as Roy Merricks and his wife Karen from MTA Travel, in Brisbane, Australia, had flown in as newly appointed agents. Roy said: 'People don't believe you when you tell them about going to space and it takes them aback a little at first. They don't realise they can book this experience now.' Another travel agency excited by the prospect was New York-based Virtuoso. Its bearded CEO, Matt Upchurch, was standing proudly wearing his Virgin Eye lapel badge. He was shadowed every step by his PR director Misty Ewing, an attentive and stunning woman who stood out in her Virgin-red tailored suit. This was an important relationship for Virgin Galactic. Virtuoso has more than 6,000 travel consultants in 22 countries, and annual sales of $3.8 bn. 'Our members have never shied away from ground-breaking ideas: that's what keeps us ahead of the game,' said Matt.

There were now cameras mounted on the podium facing the stage, with technicians and film crews working around the gathering and setting up satellite links. Princess Noor of Jordan was being offered coffee and croissants, and there was also a smattering of Virgin Galactic's Founding Astronauts, many with their partners. Jon Goodwin and his wife Pauline were delighted to be at the event, having come over from England.

'It's a fantastic scene here today. I mean, we have Buzz Aldrin and Richard Branson, just chatting together and sitting there on the stage. I just thought that was all amazing,' he said later. His wife, Pauline, nodded in agreement. 'I've only ever seen Richard Branson once before. He was coming across the Atlantic in a speedboat. His first port of call was St Mary's in the Isles of Scilly and we were on holiday in Tresco, so we made our way over to welcome him home. But being a part of this is extra special. I haven't met him yet.'

The whole event was being set up for the London media – to hit the lunchtime news. Sky News, BBC and ITN were all out in force, ready to deliver live broadcasts to their one o'clock shows. But first the reveal, the piece of theatre Miles Peckham had been planning. What happened next would have pleased Andrew Lloyd Webber.

Will Whitehorn, wearing a dark-blue Virgin Galactic T-shirt, bounded up onto the stage. The murmur subsided and there was a hush in the vast hall. 'Good morning, ladies and gentlemen. Thank you for coming here so early this morning.'

Without messing, Whitehorn set out his stall. 'NextFest is an annual event and is sponsored by *Wired* magazine. And one of the great things about it is that it showcases the technologies of the future. Technologies that are going to be very necessary if we are to win the battle against climate change and global warming in the next fifty years. We are here today from Virgin Galactic to give you a little sneak preview of the future of our project,' he boomed confidentially.

Whitehorn then recapped the story so far. 'In the 1990s, when Peter Diamandis put the X Prize together, the idea was to create a prize in the private sector for new ideas about how access to space could happen in the future – and be affordable for all. And also how new technologies and new ideas could be brought to the fore, different to those being used by the big space agencies, whose main role in space is exploration and the science of space.'

He said Virgin Galactic had been very attracted by one particular project when the company was created in 1999, and told how Virgin had been working with Burt Rutan at Scaled Composites to build a lightweight aircraft as a showcase for environmentally friendly technology. This was the Virgin Atlantic GlobalFlyer aircraft flown by Steve Fossett.

'We got to know Burt quite well – and he showed us a little project he was working on called SpaceShipOne. The rest is history since then.

'What we are going to be doing today is show you an exact replica of SpaceShipOne and giving you a preview of the size and scale of what we are currently constructing in the Mojave

Desert – SpaceShipTwo. You will see an interior cabin that will give you a very good idea of the size of the cabin we are building and it's very close to what is under construction at the moment.'

Whitehorn continued his presentation by explaining that Virgin Galactic was building an eight-seater vehicle, with room for two pilot astronauts and six paying passenger astronauts. He explained that the spaceship will be launched from an aircraft, WhiteKnightTwo.

'They are both carbon composite and both highly efficient. They use much, much less energy than any other method of getting into space so far.'

He explained that Virgin would not be using ground-based rocketry, 'which is environmentally unfriendly' and carried on, 'We are carrying our spacecraft up to 60,000 feet on the WhiteKnightTwo aircraft and launching it from that altitude. The really exciting thing about the project is the economics of it. For the first time we are going to be able to take you into space, to see the beauty of this planet and the curvature of the Earth. And you will see that thin blue atmosphere for amounts of money far less than ever before.

'Our initial tickets are $200,000 and our Founders and Pioneers are helping us to fund the project by coming on board at this stage – and we have nearly two hundred now. We hope, though, over the next five to six years to get the prices down, so that more people can afford to do this. And they themselves will help to take this project forward.'

Whitehorn was now in full flow. His eyes filled with the vision for what might happen next.

'The idea of the SpaceShipTwo project is that we will hopefully build and develop a SpaceShipThree out of it. And that will allow us to take science and payload into space.'

Then he posed a rhetorical question. 'And why does this matter? Well, organisations like NASA have a great and honourable history. They have been at the forefront of the exploration of space. NASA is also at the forefront of telling us about what is happening to the planet now.'

The Virgin Galactic chief was switching temporarily into lecturing mode. Whitehorn's brilliance as an off-the-cuff speaker without notes takes him into some uncharted and controversial areas, but he doesn't care.

'The debate about climate change and global warming wouldn't have reached the sophistication that it now has, without organisations like [NASA's] Goddard Center . . . who have been telling us for years about what has been happening in the atmosphere because they've seen it from space. But space matters for more reasons than that. Without space, we couldn't technologically tackle the issues that we now face in the world. We're going to need things like the GPS system and what replaced it in order to get the logistics of this planet working in an environmentally efficient way.'

Then Whitehorn allowed a little of his Scottish Presbyterian background to emerge. He issued the call to arms that has become a recurrent theme in his philosophical standpoint to justify space tourism. 'And the other thing is, that if mankind is going to have a hope of civilisation in the future, we've got to be able to dream. We've got to be able to believe that we can go beyond this planet and find other ones in the future – or we won't look after this planet properly. If we are told that we can never go into space and human beings don't belong in space, then frankly, we don't have a future on this planet either.'

Just as he appeared to be delivering a sermon, a switch in his brain told him to move on. 'Anyway, enough of the sillier stuff. I'd like to introduce you to Sir Richard Branson. Thank you very much.' There was a ripple of applause while Branson bounded up onto the stage.

'Thanks for coming. It's a particular honour to have Buzz Aldrin here.' More applause echoed around the huge area. 'As you all know, he enjoyed a number of magnificent space flights, including the most historic.'

Branson was more pensive and nervous than Whitehorn, with more ums and ahs. There was no fluency to begin with. He said that *Wired* had put Virgin Galactic in the Green area of the exhibition hall along with renewable energies, 'which might seem a little strange for space travel. When a NASA spaceship takes off . . . the energy it puts out could power New York.'

But Branson said the Virgin Galactic fuels team will be able to put SpaceShipTwo into space for less environmental damage than one Business Class Virgin ticket across the Atlantic – with almost no CO_2 emissions at all. 'And we hope by the time we fly regularly, that we might be able to get rid of these CO_2 emissions as well.'

Branson started to move into his stride. 'So what we are talking about is allowing people to have the magnificent experience of exploring space without any detrimental impact. It's going to be very, very exciting.'

Then he turned to announce the reveal. 'If you will bear with me, I will unveil the Virgin Galactic spaceship. One, two, three.'

It was just after 7 a.m. New York time and the techno music blasted out with lights dancing over the spacecraft as the curtains pulled apart, inch by inch, giving the eyes a few minutes to feast on the craft, like a vast stubby ballpoint pen that has been clicked off.

Stephen Attenborough, now head of Astronaut Relations, then jumped up to join Branson and Whitehorn. Buzz, sitting in the front row, appeared rather grumpy and muttered to no one in particular, 'You won't get me up in that thing.'

Attenborough, a youthful and glowingly healthy executive, was a refreshingly fresh-faced contrast next to the peroxide and wrinkles of Sir Richard and the tired and puffy-faced Will Whitehorn. Stephen had none of the Whitehorn swagger but he displayed a genuine belief in the project.

'With the very honourable exception of Buzz Aldrin, I think taking the ride on SpaceShipTwo will be unlike anything anybody in this room has ever done,' he said. 'Today is about starting to properly describe that experience. The conceptual mock-up behind me is the first time we have been able to bring some tangibility to the description of the vehicle that will be taking you into space.'

Stephen then unveiled an animation of the flight of Space-ShipTwo, which was soon to become one of Virgin Galactic's favourite marketing tools, enjoyed by millions.

'Both the interior mock-up and the animation have been

designed for us by a hugely talented team at Seymourpowell, under the creative direction of Philippe Starck.'

He revealed that 65,000 people across the world had already registered as future astronauts. 'Demand for a personal space experience continues to be really strong and to meet this we have appointed some of the world's first accredited space travel agents – first in Australia and now in America with an exclusive deal with Virtuoso. And I can reveal that Virtuoso's CEO Matt Upchurch likes the product so much he has decided to make a reservation himself!'

Then it was cue the animation. And for the next five minutes it captivated the audience with its ethereal quality and its oblique references to Stanley Kubrick's *2001: A Space Odyssey*. But sophisticated digital puppetry wouldn't be enough to make the lunchtime bulletins back in London. The big news corporations all required some real human interest and Branson jumped back on stage for the next act.

'One passenger who has flown on Virgin Atlantic for quite some time now and has collected two million Virgin Atlantic miles is called Alan Watts. And I would like to shake his hand because he is going into space,' smiled Sir Richard.

There were a few gasps, cheers and then a thunderous round of applause. But Alan, a 51-year-old former electrician from Harrow, Middlesex, who was now the managing director of an electrical engineering firm, was a very shy interviewee. At first, he didn't even want to get on the stage – but he was soon being fully baptised in the Virgin experience.

Here was a normal guy redeeming his two million Virgin Atlantic Flying Club miles for a trip into space – with the help of his wife, Heather, who had claimed miles on her Virgin credit card. 'The wife just spends a lot,' he told reporters. This was the story the UK national tabloids wanted. The only problem was that the *Evening Standard* had already scooped the story in its first editions. Jackie McQuillan had been working her devious magic and when the tabloid writers in New York phoned their news editor they were sent off with a flea in their ear to get something fresh. The good thing was, there was plenty more to say.

19. A FOUNDER WITH A MISSION

The loving couple – wrapped in fluffy towels – sat giggling and laughing on the striped couch in their San Diego home. Moments earlier they had been frolicking in the swimming pool in their back yard, before sprinting back into the house and flopping onto the seats. Jo, a photographer, and his wife Lina, a bio-scientist, began sipping glasses of chilled Napa Valley Chardonnay and shovelling in mouthfuls of Pacific prawns as the television flickered in front of their pine coffee table. They'd both finished hectic days at work and were now happy just to chill out together.

Lina Borozdina is a beautiful woman: tall and striking with a traumatic life story etched in her bright green eyes. She was born in Odessa on the Black Sea in 1969 and her father was a KGB spy who fell out of favour with Leonid Brezhnev in 1979 over the invasion of Afghanistan. Jo is a well-packed guy, with short-cut greying hair and sailor's sunglasses tied around his neck on a string. He is a laid-back racing yachtsman, born in Scotland, who is devoted to Lina.

As they sat flicking through the channels, Jo stumbled on something that caught his eye. He recognised the bearded figure of Sir Richard Branson. The programme was called *Rebel Billionaire* and the contestants were being challenged to come up with an advertising campaign for going up into space.

'Hold that,' said Lina. 'That looks interesting. What's that about space?'

'That's the British guy, Branson,' said Jo. 'He owns the airline and he's done some sailing adventures – stuff with balloons.'

'What's it all about, then?' asked Lina.

As the show – broadcast in December 2004 – unfolded the couple became absorbed. Branson was encouraging the reality-show contestants – some freaking out with laughter as they enjoyed weightlessness on the Vomit Comet – but what was more fascinating for Lina was he said he was planning to take paying passengers into space – and very soon.

Jo recalls the sequences of events. 'I have been a fan of Richard Branson from way-back-when because of my interest in boats and anything on water. I recall his Virgin transatlantic racing day and the high-altitude ballooning. I've always admired his sense of adventure.'

Also featured in the programme was Burt Rutan's company and SpaceShipOne. 'I had known about Scaled Composites for a while because of my connections with sailboat racing. I was always interested to know what they were doing with carbon fibre,' he says.

But Jo had Lina's interests at heart. 'I have always known about Lina's dream to go into space and I guess it was her turn to do something really exciting. She's supported me over the years in my sailing. Now, it was her turn.'

Lina's life had not been easy until she met Jo nearly ten years earlier. As a girl of three her father tucked her up at night, telling her a bedtime story about sneaking onto a spaceship and shooting up into space with her cousin. And, from this beginning, she always wanted to be a cosmonaut. Virgin Galactic could make that dream come true.

The couple live in a one-storey home in Pacific Beach, a seaside suburb of San Diego that has become increasingly gentrified; real-estate prices have risen sharply. It's not quite a sea view but if you squint your eyes it's possible to see the ocean, half a mile to the west. The house has a large garage, front yard with tropical trees and a back yard with its pool, a patio, balcony and barbecue. There's an extra floor at the back that has been converted from a two-roomed granny flat to rented accommodation. It is this house that is being mortgaged to pay for Lina's ticket to go into space.

As she sat watching Branson's show with Jo, she was unaware how her life was going to take another twist – this time for the better. A few weeks later, on 6 February 2005, the couple were watching Super Bowl XXXIX between the New England Patriots and the Philadelphia Eagles when an automobile advert caught their eyes. In a thirty-second advert, Volvo unveiled its XC90 by comparing its power to a rocket blasting into space. At the end of the commercial, the ship's pilot revealed himself as Sir Richard Branson.

Virgin had been approached by Volvo to appear in its prime-time slot on the Super Bowl – their biggest advert of the year. Volvo wanted viewers to give their new SUV a test drive. Virgin Galactic agreed that test drivers of the XC90 would be put into a draw for a ticket into space. The response surpassed Volvo and Virgin's expectation. Virgin had to put on a new server for the website, and it was Volvo's most successful campaign ever.

Remembering how much they had enjoyed the series, Jo went onto his PC in the study later and sent an email to Virgin telling them why Lina was an ideal candidate. He kept it a secret and fired it off, then swiftly forgot about it.

The response to the Volvo ad was overwhelming. Several weeks later, while sifting through thousands of other emails, a Virgin Galactic marketing executive opened Jo's in London and read it with interest. She decided to contact California immediately.

'My initial call was to Jo,' remembers Susan Newsam, Virgin Galactic's director of marketing. 'We were very intrigued by the application.'

His email had been to the point:

Dear Virgin Galactic.
   My wife Lina was born in Russia and wanted to be a cosmonaut when she was young. But she had to leave her homeland and now lives in California. She is passionate about space but scared of flying. She would even mortgage her house for a ticket. She wants to go. How soon can you fix it for her? She has to fly with Virgin Galactic.

'I knew when I read his email that this was someone who definitely wanted to fly with us,' said Susan.

Jo answered the call from Newsam at 3 a.m. local time. When Susan apologised for waking them up, a bleary-eyed Jo at first thought it was one of his sailing chums playing a practical joke.

Jo was curious about how much it would cost. He wanted to know what the deal would be.

'After the calls, I arranged to meet them in Los Angeles,' says Susan.

Lina's was a remarkable story that touched a lot of the human-interest buttons for a company preparing to take passengers into space – a former Soviet woman, now living in California, who would mortgage her own home to buy a ticket into space. Not only that, she was photogenic, articulate and confident. Her own tale is worth recounting.

'My father spent a lot of time in Asia, the Middle East, India and he was in Burma for a few years. He was in the KGB and travelled a lot but the Middle East was his territory. As a kid, when we lived in exile in Kyrgyzstan for four years, the phone would ring in the middle of the night and my dad would disappear for three or four days. I wrote an essay in school in second grade saying: "My dad's bosses obviously don't like him very much because they make him work at the weekends."' Her parents got called into the school where some Soviet officials chastised them for speaking to their children out of turn. From six to ten years old, Lina grew up on a grim military base. 'We had a black-and-white television with two channels. The only time there was cartoons was between 8.45 and 9 p.m. and they would say: "Good night, kids". It was five minutes of puppets and ten minutes of cartoons and I loved it.'

But Lina's escape was reading books and she was a keen pupil. 'I suppose I got the right combination of genetics from my mum and dad,' she says. Her mother was an interpreter with a degree in English language – and the Ukrainian table tennis champion. She was Jewish and born on the run from the Nazis as they surrounded Odessa, on the Black Sea. Urii, her father, a polyglot who spoke twelve languages, came originally from Moscow and was sent to Odessa while working in the secret service. He met Emilia and they were married inside two months. Lina was born a year later in 1969.

'He taught me English and when I was a kid he also tried to teach me French and German as well.' But her dad's spaceship stories became a real fascination and led her to develop an encyclopaedic interest; she could recite the names of all her Russian space heroes and through the years she has kept the childhood storybook about floating around in space.

'I grew up with the backdrop of the Space Age. I always wanted to be a cosmonaut. My parents once told me there is nothing that you cannot do in life if you try hard. My dad has been very supportive in my life. When my mum got sick, he was both my mum and dad.'

But the family situation deteriorated as Urii was ordered to go to Afghanistan. 'My father resigned when I was ten years old because he didn't want to serve the mandatory two years in Afghanistan in 1979. We were just right on the border and I was with the kids the day the war started and I expected bombs to drop on our heads. We were very scared and for a whole year I was afraid to go to sleep.'

There was a period of fear and uncertainty. Her father's boss and colleague was hardline Communist Party leader Yuri Andropov, head of the KGB who briefly became president of the USSR in 1982 before dying in mysterious circumstances sixteen months later in February 1984.

President Brezhnev denied his request to move back to Moscow. So Lina, with her mum and dad, moved back to Odessa to share with her mum's family in an eight-family communal flat, with eight stoves, two sinks, one wooden box toilet and no hot water. 'We had one large room, which was divided in two.'

Life in Odessa, a city of a million where the first Russian revolution began, was actually more pleasant for a teenage Lina than in Moscow. The weather was milder, there were lots of friends and everyone seemed to know each other, she recalls. 'It is a beautiful place on the Black Sea. I was lucky to grow up in such a gorgeous place.' Not only that, but she was doing very well at school, especially in science. Lina was also good at shooting, taking part in national competitions, and swimming, although she doesn't share Jo's passion for competitive sailing. 'I stay on shore waving the flag and acting as support crew,' she says.

Lina's mum died when she was thirteen. 'My dad and my mum's parents then raised me.' In 1991, when she was 22, she travelled to America with thirty unconvertible roubles in her pocket. 'It wasn't easy getting out of Russia. I came for a guest visa and applied for political asylum. There were some things happening back home and I had to leave.'

Lina is vague about the reasons but admits she was beaten up badly and implies she was a victim of rising anti-Semitism, although she is not Jewish herself and her father is a patriotic Russian. 'It was something to do with my nose,' she says, pointing at her Roman curved appendage. 'It is something like my father's but my father is a pure-blood Russian. People have approached him recently looking at his heraldic past. Maybe I'm a long-lost Russian princess,' she jokes dryly.

She came to southern California and went to stay with her mother's sister, Eva, and her family in Los Angeles, then applied for asylum. Her close-knit relations helped guide her through the INS system, filing papers and dealing with the culture shock of arriving in the land of plenty. Lina was denied asylum at first and she initially married a Russian guy, but she was a stateless person for thirteen years. Meanwhile, she took a job at five dollars an hour washing dishes in a biotech company. It took her eight years to get a green card and she was only granted US citizenship in 2004. 'It took a very long time. It's been very trying. The first thing I did was get out on a plane to the Ukraine to see my dad, stepmom and old friends in Odessa,' she says.

'At last, I felt I belonged somewhere and being in limbo I was afraid to go anywhere – or over the border into Mexico which is fifteen minutes away – in case they didn't let me back in.'

Love blossomed for her and Jo. They met after she went into his Pacific Beach studio and asked him to take some Valentine's Day glamour shots for a previous boyfriend. They became friends and Lina did some modelling for some clothes stores. They hung out together for a while and married on 28 August 1999.

Armed with an MSc in biochemistry, her science career was moving up and she worked as a chemist in San Diego in several drug technology start-ups, where she was becoming an expert in the creation of synthetic DNA.

'I'm a nerd. I work in the lab with goggles and a white lab coat. I mix stuff up in a tube and blow things up. It's a fascinating subject and I love what I do for a living.'

But now she could become one of the first space tourists.

'I really want to get on one of the first flights. Originally there was only going to be seven women and now there are about twenty,' she says.

'I'm the only one who has mortgaged my house. If I can do it, then so can a lot of regular people. You don't have to be millionaires to go into space. I know $200,000 is a lot of money, but I did my budget. Jo and I live a pretty Spartan lifestyle, we don't have any debts and we don't have any credit cards.'

Lina's name was now in the lottery hat with nearly eighty other founding astronauts and a lot was happening in London to look after this new group. The arrival of the new marketing director Susan Newsam meant Stephen Attenborough now had someone to pick up more interesting leads. So Newsam flew out to meet Lina and Jo at the Virgin Entertainment offices in Los Angeles. 'I was thrilled, probably from my own point of view because it was another female,' recalls Newsam. 'I thought she was bright, intelligent and beautiful. She was passionate about the project and about space. I thought what a fantastic ambassador she would make. And Jo was so supportive.'

Susan Newsam explains that those early customers were direction-finders. Not only in terms of their commitment with money but for fully appreciating what the astronauts wanted and expected for their $200,000. 'We now had the time to spend with them to find out really what they wanted as well. This helped us in the early design of the project.'

In Malibu, further up California's Pacific Coast, another Virgin Galactic Founder would not have to countenance selling her beach-front home to buy a ticket. Victoria Principal is a multi-millionaire business founder who heads up her skincare company, Principal Secrets. It is part of the successful Guthy-Renker empire where celebrities sell endorsed products direct to the public promoted by *infomercials* on a cable television channel. It has sales of $500 million a year, and Victoria is one of the star turns.

Victoria Principal, now 60 and recently divorced from her plastic surgeon husband, is an extremely smart lady who once studied to be a commercial lawyer before her acting break-through. She remains the glamorous icon of the hit series *Dallas*, the big-haired Pam Ewing who captivated millions in a tale about the oil-wealthy Texans.

'After *Dallas*, I started the business in 1989 and that's a full-time job. I continue to be the CEO and president of the company,' she explains.

Her white-washed villa with its verandahs, bougainvillea and courtyards is a sumptuous designer pad, immaculately decorated and fastidiously ordered. A few steps from the balcony leads down to the white sands and a nearby pier. The sky is azure while the surf gently laps the shore.

'It was extraordinarily beautiful last night with the phosphorus in the ocean and a thousand stars dancing on the crest of each wave,' says Victoria, poetically.

Each morning she wakes at 6 a.m., takes a walk along the beach and starts her day watching the sun come up. 'I usually do yoga, have breakfast and I pinch myself everyday,' she admits. Then she heads to the office in Ocean Park Boulevard, Santa Monica, while three days a week she comes home to have lunch in Malibu. Every Friday she takes off and heads to a remote cliff-top cabin or goes travelling. But her new-found passion is for Virgin Galactic's project, and she has been asked to help Richard Branson personally.

'My enthusiasm for this is intense. I've always enjoyed adventure which coincides with speed.' Victoria was born in Japan in 1950 when her father, Victor, was an US air force sergeant but an accident prevented him becoming a pilot. 'He certainly passed on his love of speed to me. I have a Formula Ford licence and I've been racing cars for over thirty years. I trained with a paraglider and I jumped off one of the highest mountains in Switzerland in 1990.'

She has also been down the Salt Lake City bobsleigh run with the medal-winning US Olympic team, experiencing 4 Gs in the terrifyingly icy bends and turning in a respectable time just two seconds outside the gold medal time.

Her craving for thrills led to her hot-air ballooning over the desert and finding one of the only two restored P59s jet fighters left in America and then doing aerobatic barrel rolls over the coast of Malibu.

When she was watching the television like Joe and Lina she was immediately hooked by Burt Rutan's vision. Ten years earlier Victoria had arranged to go up in a Russian MiG. 'At that time, taking a trip to Russia was more dangerous than the flight up into the air.

'When I saw the programme, I had a pretty good idea of what was involved because I had been interested in space flight. I learned a lot from my earlier involvement, but when I phoned Stephen it was obvious to me that Virgin Galactic's programme was so much more advanced and thorough – and much less invasive to the atmosphere.'

Sir Richard Branson jetted over to meet the former TV star and convinced her to become actively involved. 'I have met some of the other astronauts and I have been made an ambassador for space. I have spent my spare time learning about the programme so I can contribute any ideas.'

Was she inspired by the Apollo missions? 'Absolutely, I recall it vividly. I was living in France at the time, training as a ballet dancer, and I got up at 2 a.m. to watch. It is burned into my memory. At that moment, I didn't think "I want to go up into space." It all seemed surreal at the time, but about fifteen years ago it occurred to me that perhaps I could. And I think a lot of people going up into space are very like-minded. I don't think that my desire to go into space is extraordinary – or that it sets me apart from the others who want to go up. I think many of us really want to see this fragile planet. This is my home and the idea of being able to see it from space just fills me with gratitude. And, on top of this, to experience 4–6 G forces – then I'm in.'

She has viewed the footage from Brian Binnie and Mike Melvill's flights. 'The only thing that I can't predict is how it will feel to be weightless and what it will feel like looking out of the window and seeing part of the planet,' she says, looking fit and trim and a walking advert for her skincare products.

Victoria wants more than a three-day training session though. 'I am going to choose my own astronaut schedule because I want to be ready for the trip. I am going to do an extra two-week programme because I like to challenge myself.'

Will that include the dreaded centrifuge? 'Yes,' she says. 'I consider it an extraordinary privilege that I am going to get to go up into space.'

Sir Richard and Victoria Principal are now concocting further plans to go beyond orbital space. 'I think we have an opportunity to treat space and space exploration more kindly and with greater responsibility than we treated this planet. I do think that we need to become more responsible for this planet. I don't think that we will be wiped out by war. I think the planet is going to reclaim itself. It will rid itself of us, unless we start doing the right thing. So in terms of going into space to find another place to live because we've wrecked this planet, I think it would make more sense to put this planet back in the kind of shape that it should be. Then space exploration can be an extension of the way we live – and the way we care for the places we live,' she concludes.

Although asking the astronauts, such as Victoria, is helpful, it was never going to be comprehensive enough, so Susan Newsam commissioned more research to see where the main market was. She found that in the United States alone there were enough seriously wealthy people who might want to spend some leisure time in space to sustain the market. In 2006, there were 8.2 m households with over one million dollars in assets, not including their homes – almost 7 per cent of the population.

'Over the last ten years, there have been a lot of surveys, including Futron's. I took all that material and wrote a report on that and then after that we also did our own survey,' explains Newsam.

The Futron Corporation had asked the basic question of whether people would like to go into suborbital space. Janice Starzyk, the programme manager for Space and Telecommunications, has become a spokesperson for this organisation. More than 60 per cent of their poll said people wanted to view the Earth from space. 'This is really great news for the industry because if someone were to go up in New Mexico they might also want to go up in Europe, Asia or somewhere else. Seeing New Mexico and Oklahoma from space is very different from seeing Sweden, the North Pole or Europe. Our research showed that was a very big driver for those who want to take the trip,' says Starzyk.

Futron also found out that, of the people they polled who said they would like to go in space, a lot wanted to experience a rocket launch – so they could say they were one of the first to do it. What price, then, for the bragging rights at the wine bar, golf club or hairdresser's to say that you were the first hairstylist from Milton Keynes to go into space?

'This has to be understood when you are looking at space travel later on. How this will have an impact on the novelty value after hundreds of tourists have gone into space.'

She concludes that the price would have to come down fairly quickly – with a figure of £20,000 putting it in the range of many more people.

But there were some contradictory results too. 'The experience of weightlessness is surprising, though; 42 per cent of the Futron survey says the experience of weightlessness is not important. I would have thought this would have been one of the coolest things. But, according to our research, it didn't seem to be as big a driver as we would have thought,' adds Ms Starzyk.

But Futron expects demand to increase from 503 passengers in 2006, to 1,330 passengers in 2010, up to 15,712 by the year 2021. While polls by the news channel CNN and the US Department of Commerce showed 75 per cent of people said they would take a ride into space if given the chance.

This was all grist to the mill. But Virgin Galactic needed to find out what people were prepared to pay and what they were looking for in a suborbital flight into space. 'The number-one thing that came out was that people wanted a window seat, they wanted to see the curvature of the Earth and, despite the early Futron findings, they wanted to experience Zero G – weightlessness,' says Susan Newsam. Armed with this information, Virgin Galactic could instruct the designers and builders more clearly. And Lina – now an enthusiastic ambassador – and Victoria would be on hand to help with the process.

20. NORMAL PEOPLE TAKE FLIGHT

Not all of the first Virgin Galactic astronauts have such dramatic tales to tell. Some live more sedentary lives. 'We're just normal people,' says Christine Easterfield, now in her early forties, who was born and brought up in Newton-le-Willows, near St Helens, in the northwest of England. Just normal people, it has to be said, with a rather healthy bank balance.

Christine, who graduated from the University of Hull with a computer science degree, is also a Virgin Galactic founding astronaut. She worked full-time in Cambridge as an IT manager in a company that is part of a multinational corporation, and recently went back to university to study English. She and husband Mark are set to become the first married British couple to fly into space together. They don't have any children, and Christine is unfazed by the prospect of going into space.

'Our names are down and we're excited about the prospects. We don't know when we will be flying, probably during the summer of 2009. There will a lottery to see which flight we'll be on – but it's written into the contract that we will be going together.'

Instead of the sun-kissed beach in the Caribbean, this will certainly be something different for the Easterfields. However early they fly, though, they won't be the first couple in space. Back in September 1992, Mark Lee and Jan Davis were the first spouses to fly together on the *Endeavour* orbiter. To NASA's dismay, the pairing of the astronauts – who married after being assigned their mission – brought a frenzy of tabloid interest about whether the couple would be the first to have sex in space. But with five other crew members in close proximity and micro-gravity at work, Mark and Jan didn't have much chance. NASA made sure they were both too busy by putting them on different shifts.

The Easterfields live in a comfortable and charming de-tached home in Ditton Fen on the outskirts of Cambridge. They have a rambling garden with apple and pear trees, oaks and elms, a croquet lawn and an Aga in the kitchen where the orange Le Creuset pots simmer with home-grown carrots and chicken. They are indeed a very ordinary English couple with aspirations. They are not ostentatious. Both are stalwarts of the local Waterbeach amateur dramatic society, where Mark designs the lighting and Christine stage-manages shows and occasionally treads the boards.

They enjoy making cider and plum jam from fruit gathered in their garden. They love English country pubs, hoppy beer at the nearby Plough Inn, and the local tradition of boat bumping on the Cam. They enjoy their holidays together, having ventured to the Antarctic on a cruise and New Zealand and they've explored much of Europe. Not quite Right Stuff material – more just Ordinary Stuff. Now space beckons. They are part of an elite group who have paid their deposits for the privilege of becoming space tourists.

Mark and Christine both worked for the same business, Smallworld, an energy company that was purchased by GE Energy in 2000. As one of the ten original founders, Mark did extremely well when it was bought out; Christine cashed in her options too. But neither had any desire to change a lifestyle they both love. Mark still enjoys going into work as a database programmer – his only concession is a Monday off.

Mark was brought up with science and engineering as his British backbone. Born in 1953, he is the first generation to have been captivated by the race into space. Like so many post-war baby-boomers, Mark was fired up by the early space adventures. The optimism of the world during the early 1960s showed few bounds. He grew up in Golders Green in London and his paternal grandfather, whom he never met, was a chemistry professor in New Zealand, while his mum's father, Ernest Braddock, lived with them in their rambling house. Ernest was an inspiration, although he died when Mark was just eleven.

'I remember he had a Hornby train set in his room and he built it all himself. The power supply, lighting, signalling and all the models. It was quite something – he was very practical and that had a huge influence on me as a boy. My father was an extremely bright man who studied maths and chemistry at Cambridge. Dad was very scientific in a mathematical sense, but I learned about the practical electrical things from Ernest.'

'I was very into space from an early age. I had Airfix and Revell kits of rockets and I took photographs off the television. I was fascinated by the Americans going to the Moon. I remember being captivated by the first space walk. On the BBC at the time we had people such as James Burke and Patrick Moore who made it real.'

These were two of the broadcasting triumvirate of the 1960s and 70s who brought space into the front rooms of ordinary people across the UK. The third was Raymond Baxter, the former Spitfire ace who made a daring low-level raid on the headquarters of the V2 rocket scientists in March 1945, who became the host presenter of the BBC's *Tomorrow's World* in 1966. After tea on Thursday evenings was when a gentle and courteous Baxter would introduce future technology gadgetry to an often doubtful nation still sitting in rooms with a single 40-watt electric bulb and coal fires. Baxter forecast the pocket calculator, the video machine, the credit card and the microwave oven. But, more importantly, he placed in many minds a sense of optimism about the future of space. He died on 15 September 2006, aged 84, but his wonderfully sonorous speaking voice and well-articulated English remain an indelible memory for Britain's baby-boomer generation.

Mark Easterfield shares a common culture with the likes of Sir Richard Branson, Virgin Galactic president Will White-horn and scores of other baby-boomers in Britain, weaned on an interchangeable mixture of science fiction and fact. The first post-war television generation watched a black-and-white box packed with science series that excited the mind. William Hartnell was the first Doctor Who, appearing on the BBC on 23 November 1963, the day after the assassination of President Kennedy. It was intended purely as a children's sci-fi drama, but the London police box, the TARDIS, became a national icon, whisking the Time Lord off to face enemies around the universe, including the Daleks. A few years later *Star Trek* would captivate British viewers, introducing warp-factor, teletransportation, anti-matter and force field to the lexicon. The commercial television channels also brought science fiction into the living room. In 1960, the impresario and television director Lew Grade met Gerry Anderson, a former Pinewood Studio film editor who had formed his own television company, AP Films, based in Slough. His children's puppet show was successful but his new offering, *Supercar*, had been rejected. His business was on the verge of collapse, but Grade took the series after cutting the budget. Anderson's innovative Supermarionation puppetry technique was a winner, and the 26 half-hour episodes were sold to America. From this came a slew of children's programmes, *Fireball XL5*, *Stingray* and then the series that defined the 1960s, *Thunderbirds*. While the world was watching the real-time exploits of the Soviet and American astronauts in space, the early-evening schedules and Sunday afternoons were filled with the exploits of International Rescue, then *Captain Scarlett and the Mys-terons* in 1967, and *Joe 90* a year later.

As this audience grew older, the arrival of new series, such as *Star Trek* and *UFO* – the first fully fledged live-action science-fiction series – continued the theme of space. Like millions of other English teenage boys, Mark was given permission to stay up late by his mother to see and hear the flickering black-and-white images on television of the Moon-walk.

'I was particularly fascinated by space. I collected all the magazines and books and I was especially interested in the Saturn 5 Rocket. The great thing about rockets was the sheer brute force. It took enormous balls to make this thing take off. The rocket never went wrong. It was a staggering event to watch and think about. And there were human beings sitting right on the top of this powerful missile,' he says.

He recalls the classic image of the rocket lifting off from the Cape Kennedy launch pad, with the first-stage booster dropping off and falling away back into the blue ocean fifty miles below. 'It was so visual with the amazing imagery of the white rocket with the American Stars and Stripes on the side.'

But it was Stanley Kubrick's glorious technicolour vision of space, with the von Braun wheel and the universe in the 1968 film *2001: A Space Odyssey* that enthralled so many. It remains one of the greatest artistic films ever made, a visual masterpiece, with its ground-breaking screenplay written by Arthur C Clarke. The photography and direction were brilliant but the film's meaning has always been rather obscure.

However, the film was a feast that captured the brilliant balletic movement of weightlessness with music from Johann Strauss's *Blue Danube* and Richard Strauss's *Also Sprach Zarathustra*. This single movie had a seminal influence on many of those still inspired by space. Brian Binnie, the SpaceShipOne pilot, recalls going to see the film in London with his father when he first travelled to Britain in the late 1960s, while Mark remembers: 'I went to see the film two or three times in Cinerama at the Empire in London. It was awesome. Kubrick knew how to use pictures and music.'

But, like so many others, Mark Easterfield's interest waned as space became commonplace and Britain struggled to keep the lights on during what seemed a long 1970s winter of discontent and industrial disputes. Space fell by the wayside as the more mundane demands of ordinary life took root.

'The curious thing was that the public lost interest. It peaked after Apollo 11 and then the world stopped again briefly with Apollo 13. It was a nail-biting story and brought the dangers of space back to the public consciousness again. Those astronauts were going to die – and there was a race to get them home. The drama was intense.'

Mark talks of something that is typical of many middle-aged people who have had their interest in space rekindled. 'There are three angles here: there is what you thought at the time as a young person growing up, what you can remember about it, and what you see now – in retrospect. It was exciting at the time but when you look at it now in terms of the technology, it was an amazing feat of planning.'

Then he comments on a dilemma many find difficult to fathom. Modern organisations are often unable to tackle major projects and deliver them successfully – and on time. It appears that it is a malaise of modern life in Western capitalism.

'I don't think we've got this kind of organisation any more, in any walk of life or in any discipline, the ability to make something like this happen on a major scale,' he said. This is perhaps why so many twenty- and thirty-somethings are sceptical about the fact that man has ever been to the Moon. It raised the major question: How come we can't get to the Moon now, when we were supposedly able to do it in 1969?

If technology has advanced so much – and there is no doubt that it has – how is not possible to do it now, rather than wait another ten years at least to go to the Moon?

'Some people say that we should just send unmanned probes into space if we want to do research. That's not the point for me. If you look at human endeavour it has to keep moving and keep expanding. The science is important,' says Mark.

Christine adds her thoughts. 'My feeling is that if I get a chance to go into orbit, I would say yes. But go to Mars? That would require something different and I don't think I'd be able to do this. Going to the Moon? I'm not sure. I don't have anything like the urge to go to the Moon. What I want to experience is the freedom of zero gravity – even just for a few minutes. That will be sublime.'

She adds, 'You will be doing something you cannot do in any other way. Yes, you can go up in the Vomit Comet and bang against the padded walls. But to see the curvature of the Earth and the black darkness of space, that's really what I am expecting. I like the idea of being in the first hundred people going into space.'

But both Mark and Christine Easterfield are relaxed – they are going together and they will wait for their names to be drawn out of the hat.

Jon Goodwin is another founding Virgin astronaut. He has the upright bearing and the age profile of a NASA-trained astronaut. Except that he is a retired sweet wholesaler from Staffordshire, England. Slim and fit for his 63 years, he has a neatly trimmed grey beard with clean-shaven upper lip. At six foot he might easily pass for someone in his late forties. Pauline, his wife, is a smaller, more petite lady, who nonetheless shares Jon's passion for sport and taste for adventure. Pauline took up motor racing with a vengeance at the age of 56. Three years later she has been hurtling her Ferrari 328 in hill-climbing races. And just short of her sixtieth birthday, she took her racing driving licence at Silverstone and began track racing. 'I am not ready to get old yet,' she insists, then repeats for emphasis, 'I am not ready to get old yet. Racing it is, and I wish I had done it before.'

So while Pauline is seeking her thrill on the racetrack, Jon has signed up as a Virgin founding astronaut to go on his own. Why isn't Pauline going too?

'We have two boys and my wife wasn't happy that, if both of us disappeared into space and we went into orbit, the two boys could never claim the insurance because of the fact that we were still up there somewhere,' he says, with a dry hint of sarcasm.

The couple live just outside Newcastle-under-Lyme at Blackbrook, near to the village of Baldwin's Gate. It's a rural location. They have been married for 35 years and have lived in this setting since then. They appear a secure and self-contained English couple. 'It's a cottage and we have every-thing we need here,' says Pauline.

Jon and Pauline (then Squire-Goodwin) both competed for Britain in the Olympic Games in 1972 in Munich and then in Montreal 1976 with the kayaking team.

Jon concedes, 'I am a bit of an adventurer, and I always have been. To be here at this time – with something totally new – I thought, the opportunity to go into space at a relatively small cost, in my opinion, was, well, I couldn't miss it.'

So did he have a passion for space from an early age? 'It isn't so much my interest in space,' says Jon. 'It's challenges that keep me going.'

He continues, 'I was the first person to canoe down the Grand Canyon in a Canadian double in 1971. Two years ago I cycled across Australia and then when this opportunity came up, it could only come for someone in this period of time; I couldn't resist going up into space.'

Jon retired and sold his wholesale sweet and tobacco business when he was 55 and he spent the next seven years until late 2006 pursuing his adventure interests. Before he goes into space he is racing in the Peking to Paris rally with his son. The six-week race replicates the original 1907 rally and a hundred vintage cars – dating from 1906 – have been granted permits and will be attempting to drive 10,000 miles, over hostile terrain, through Mongolia and across the Gobi Desert. John will be taking his Aston Martin DB6, the youngest car in the rally, from 1969.

'The car has to be totally self-sufficient, so we have spent the last two years preparing the car. Anything that goes wrong with it you have to fix yourself. We can't have a back-up team or follow-up team. My son's a civil engineer and he's a very good mechanic.'

Pauline says, 'Jon is very fit. But I think that comes from having participated at a certain level in sport. It's natural to keep fit throughout your life – it never leaves you.'

Jon was also with the party in the Mojave Desert in May 2006. 'To go around the facility with Burt Rutan, and hearing him talk about building this spaceship and everything to do with it, it was an amazing experience. Then to be taken up by Brian Binnie, the test pilot, and put through three and four Gs in a jet! It was incredible.'

There are one hundred founding astronauts from around the globe – including Xavier Gabriel Lliset, one of the most popular and celebrated men in Spain, who created La Bruixa d'Or (The gold witch) a kind of *Jim'll Fix It* with a lottery attached – and twelve from the UK.

'It's going to be interesting to see what order we come out in. Jon is hoping to be in the first batch,' says Pauline, adding that if for some reason Jon couldn't go she would take his place. 'On first thoughts, I would say I'm not really sure about this because it needs to be tried and tested – but I'm getting involved in it so much now it seems very exciting.'

While the identities of many Founders have not been disclosed, Whitehorn and Attenborough have talked openly about Adrian Reynard, a British race-car designer and expert in the use of composite materials, which are also used extensively in Virgin's spacecraft: 'It's been a great confidence-booster to all of us, and to Burt [Rutan] personally, that Adrian Reynard was one of the first people to sign up for this.'

Another of the Founders is actress-turned-entrepreneur Victoria Principal, best known as one of the stars of the 1980s TV drama *Dallas*. Whitehorn said that she will serve as an 'ambassador' for Virgin Galactic and hopes she will encourage more women to sign up.

And Trevor Beattie, the advertising wizard behind a string of successes including the high street retail chain French Connection UK, and who invented a dyslexic swearword to excite millions of young people into buying a range of clothing, has also signed up, offering his creative expertise.

But Sir Richard Branson is keen to help one unique individual experience a suborbital space flight – Professor Stephen Hawking. In December 2006, in an interview with BBC Radio Four's *Today* programme, Hawking revealed that despite his paralyzing motor neuron disease and confinement to a wheelchair, he would still love to go into space.

He also said that human survival depended on our exploration of the universe. 'It is important for the human race to spread out into space for the survival of the species. Life on Earth is at the ever-increasing risk of being wiped out by a disaster, such as sudden global warming, nuclear war, a genetically engineered virus or other dangers we have not yet thought of,' he said.

The celebrated English cosmologist revealed he had an abiding ambition to go into space. 'Maybe Richard Branson will help me,' he said in a passing comment. The radio programme was heard by Will Whitehorn and Sir Richard contacted Professor Hawking's office to find out if it this would be possible.

'We are now looking at the possibility of taking Stephen Hawking into space. Of course, it will depend on the medical condition and what he will be able to undertake – but we are very keen to take Professor Hawking into space if he gets the go-ahead,' said Whitehorn.

Hawking, born in January 1942 in Oxford, is one of the most distinguished astrophysicists of the post-war generation. He took a first class honours degrees in Natural Science at University College, Oxford, before going to Cambridge to do research in Cosmology. In 1973, Hawking joined the Department of Applied Mathematics and Theoretical Physics and since 1979 has held the post of Lucasian Professor of Mathematics, a chair founded in 1663 and previously held by Isaac Newton. For Virgin Galactic, the prospect of taking such an illustrious scientist into space would be both an honour and a massive public relations coup.

Meanwhile, the founding astronauts must wait to hear when they will be going. But the first commercial flight – after all the testing – is already spoken for: that flight will include Burt Rutan and Sir Richard Branson, as well as his mother Eve and father Ted, who will be ninety if the flight takes place as currently scheduled. Also on board will be Branson's son Sam and daughter Holly.

21. READY FOR TAKEOFF

The space tourist will be paying good money for the full-blown, sensory experience of space. But the physical punishment of blasting off in a rocket from a launch pad is unlikely ever to be a major sales tool – and it is hardly cost effective either.

Virgin Galactic is already working to ensure that the experience is not only inherently safe, but more benign and suitable for many more people. While insurers are now placing space tourism in the same bracket as extreme sports such as bungee jumping, Himalayan mountain climbing and free-fall parachuting, space tourism does not want to be seen in the category of high-risk adventure. It needs to become a mainstream tourist experience – with levels of safety matching modern air travel – if it is to attract thousands of people willing to fly.

The average space tourist will never be as fit as a professional astronaut or cosmonaut. But, according to Virgin's space-science team, this shouldn't stop around 85 per cent of people being considered for its space flights – if they undertake a proper flight programme. And with time and experience 95 per cent of the population should be able to travel.

In early 2007, a rapidly expanding Virgin Galactic flitted from its warren of offices in Half Moon Street to a much more bespoke and high-profile site in London's Leicester Square. Here, amid the neon lights, where tourists queue for cheap tickets for the capital's leading cinemas and theatres, the spaceship company has its sparkling offices. Philippe Starck's massive Virgin Galactic logo – the famous cosmic Eye – gazes down on the crowds milling around in the square.

Inside the brightly lit offices, Dr Julia Tizard sits behind her desk in a suite of private rooms, a kind of Harley Street specialist for those who want to go into space. When astronauts sign up, they are offered the option of coming in to meet Julia or one of her medical associates to discuss the trip ahead. It's a relaxed and informal way of dissipating some of the trepidation the potential voyager might feel.

Julia Tizard sounds like one of the exotic crew of the Starship *Enterprise*. An ideal foil for Jean-Luc Picard perhaps? But she is not from some far-off galaxy – she was born in Blackpool, Lancashire, and she will be responsible for determining whether Virgin Galactic astronauts can take the ride of their lives.

'It's perfectly natural for people to feel some nervousness – this is something new and exciting. But our job is to reassure people that if they follow the procedures that have been set down, then all they have to do is enjoy the trip,' she says.

It has been a meteoric career rise for Dr Tizard, who worked at the European Space Agency on satellite projects and was an intern at NASA in 2000. But space – and human physiology in zero-gravity – is her abiding interest. Armed with a PhD in astrophysics from the University of Manchester, gained after her first degree at the same university, she is at the forefront of an emerging discipline – health and medicine for space tourists.

She was in Vancouver at the World Space Congress in October 2004 with a bunch of her friends working in space-related fields. 'We realised the second and final X Prize flight was due to go off the next day. It suddenly occurred to us that we had come all the way over from England and we were on the right side of the right continent, so we made a spontaneous decision.' They jumped in a couple of taxis and drove to the airport, bought some cheap tickets and flew down to Los Angeles. They arrived at 8 p.m. and called one of their friends working at Cal-Tech. He and some friends grabbed their cars and took the Vancouver visitors out to Mojave, eighty miles from Los Angeles. 'We slept in the car overnight. I remember it was bitterly cold. We got up at 5 a.m. in the morning to watch the winning X Prize flight with Brian Binnie.' So they saw White Knight with SpaceShipOne tucked in underneath take off and begin its long climb to 45,000 ft before releasing Binnie on his historic flight.

'This was the first time I heard about Sir Richard's involvement in the project,' says Tizard.

For her, the atmosphere at the airport was electric as she joined the thousands awaiting the moment. One of her friends, Loretta Hildago, now a Virgin Galactic Founder astronaut, was working on the X Prize project and managed to secure a handful of VIP tickets for Julia and her friends. 'We stood on the platform, beside the runway, and we saw Brian Binnie get into the spaceship. We saw it take off and then two hours later we saw it land. It was incredible to think this was a breakthrough moment in space history. He had been to space and back again. It was very real.'

Julia and her friends watched the flight, grabbed some lunch and darted back to LA to get a flight back to Vancouver. 'It was funny. Nobody actually noticed we had gone when we got back – but they were all talking about the X Prize flight.'

The cosmos is in her veins. Her father, John Tizard, was a scientist who built wind tunnels to test planes, working for various British Aerospace projects all over Europe and in the United States. His work, including work for NASA, took him around the globe, so Julia, who grew up with her mother Melanie and sisters Hannah and Natalie, was schooled in Europe. After her X Prize drama, Julia was due to finish her

PhD the following year and had already made a decision that she wanted to join a commercial space company. 'I wanted to get involved with Richard Branson's new space business – and I spent the next year trying to make it happen.'

Tizard approached Virgin Galactic, flagging up the need to assess the medical condition of every one of the potential passengers. Here was a new opportunity for science opening up. She was keen to build a project around space health – and fitness for space flight. Virgin gave her a three-month trial and it has turned into a brilliant full-time job.

'We are now looking at the exact experience we want to create. Then we will build the operations around this. The space flight is the *crème de la crème*. It will be the icing on the cake of a very special trip. And we need to ensure that this experience is as good as it possibly can be, so preparation is everything and the build-up is really important,' says Tizard.

Most Virgin Galactic astronauts will arrive at the spaceport in the late afternoon or early evening, travelling initially to Mojave until Spaceport America in New Mexico is completed in 2010. Sundown at 4,000 ft in the New Mexican desert is an enchanting time of the day as the air cools and the sky turns a warm scarlet. 'By the time you arrive, you will know the name of your flight – named after constellations such as Cassiopeia, Draco, Cygnus and Cepheus – and the fellow passenger astronauts who will be flying with you. Of course, this will be on condition that they fulfil the training schedule and medical tests successfully.'

Each astronaut will have been checked out by their own family doctor to ensure they have nothing serious that might prevent them undertaking the trip. This is a preliminary stage before Tizard and her team undertake more specific tests. Those with serious heart abnormalities, coronary artery disease and diseases affecting the body's major organs will be advised not to make the flight. They will also have had some dietary advice about how to prepare for the trip.

During their first day at the spaceport, the astronauts will be split off from friends and family for their training. But their families will not be neglected – indeed, they will either get the pampered six-star treatment in the Virgin spa, a spin in a fastjet, or perhaps a local excursion to meet some space aliens in Roswell!

Under US Federal Aviation Authority instructions, every Virgin Galactic astronaut must undergo an intensive training programme lasting three days. Each person will be informed: 'We will be undertaking a series of briefings for you over the next three days so that you are clear about all the protocols of going into space. This is not like another flight. For a short period, you will be experiencing true weightlessness and we want to make sure that you gain the best possible use of this very precious time in space.'

One rival to Virgin Galactic is Rocketplane XP, which will be taking off from the Oklahoma spaceport. Their space-flight itinerary includes a Civilian Astronaut Space Training (CAST) programme lasting four days. The first two days have been developed with state and US federal agency co-operation. The space tourist will learn about aviation, physiology, flight trajectory, and take part in simulations in an altitude/hypoxia chamber and for spatial disorientation – which involves being spun around in a hi-tech chair. The third and fourth days involve preparations for the flight, meeting the crew and looking at procedures. The flight itself takes off like a conventional business jet, climbing to 20,000 ft. Then the pilot – former NASA astronaut John Herrington – switches to ignite the rocket engine, which burns for 70 seconds, accelerating the XP to more than 3,500 ft per second. At an apogee of 330,000 ft, the space tourist will enjoy three and a half minutes of weightlessness.

The Virgin Galactic experience will follow a similar path: an induction session, training and relaxation techniques, some positive fun, such as flying in fast jets, a session on the regulatory requirements, the build-up to the flight, the final countdown, the actual two-hour flight, coming back to Earth, and then the celebratory party.

Tizard is using every avenue available to ensure that health issues do not hinder the pleasure and enjoyment of a suborbital trip. 'It depends how well our Founder astronauts do in their initial training,' she says. 'But the point of our business is to be able to take anyone who wants to go at three days' notice – so they don't need to have six months of astronaut's training.'

So will they be whirled around in the centrifuge? This is a hardcore piece of machinery, verging on an instrument of torture. Most cosmonauts and astronauts come to dread prolonged sessions being spun around at pressures of up to 8 Gs, where they would often pass out of consciousness. Surely this isn't a way to enjoy a trip into space?

The TsF-18 centrifuge at Star City in Russia is the largest in the world and can generate up to 10 Gs for manned training. You are recommended only to take a light meal two hours before climbing into the cabin – where your heart rate, breathing and other vital organs are measured and monitored. You must keep your limbs loose, finding a comfortable position. This is a solitary and scary experience as the giant arms begin to rotate. Breathing becomes difficult and, to reduce nausea, space tourists have to breathe from their abdomen, not the chest. If you hold your breath, you will quickly pass out. But you have the ability to stop the experiment at anytime – using the dead man's handle. If it becomes unbearable, you will have the opportunity to end the agony.

'If our Founders find the centrifuge is a real help when they are training, then we will keep doing it,' says Tizard. 'There is a balance between doing the fun stuff and making it a positive preparation for the flight. We want to make sure that we are covering everything we need to.'

She adds that, for many paying passengers, a spin in the centrifuge will be expected. 'They have paid for the real astronaut experience and a spin – however unpleasant – gives them an indication of how they will react. The other thing to remember is that you can program the centrifuge to whatever velocity you require. We don't have to push people as far as NASA astronauts or Star City cosmonauts. That's extreme and we don't need to do that.'

Julia, who is slim and fit, has already tried it several times at QinetiQ, formerly the UK's defence research and science agency, at Farnborough. The QinetiQ centrifuge runs up to 9 Gs. For her, it peaked at 6 Gs for a few seconds, but the average pressure on her was 4 Gs. That is not too arduous but it is more than the space shuttle generates. Its maximum is 3 Gs, but then for an astronaut who has been weightless for a few weeks or even months, this can be a punishing amount of pressure to take on the body for too long.

Steve Bennett, the chief executive of Starchaser, the English rocket company, which plans to fly from New Mexico in the next few years, has no qualms about putting his paying passengers through such an unadulterated experience. He wants his tourists to experience the actual flight of the early astronauts. 'We want people to get the same experience of Alan Shepard when he went up in the Gemini capsule. It will be straight up into space with a full spacesuit and then a return to Earth. Exactly as it was in 1961,' he said.

But Virgin Galactic expects to take thousands into suborbit, making it as easy as possible for people to have access to space. One woman who has signed up is paralysed and in a wheelchair – her ambition is to experience weightlessness so she can be temporarily free of her physical shackles. Virgin Galactic is working on a programme to make this possible for people like her and Stephen Hawking.

The wife of one of the Founder astronauts needed a little extra scientific reassurance. Dr Alan Finkel, a neuroscientist and the founder and chairman of *Cosmos* science magazine in Australia, bought four Founders tickets, including one for magazine colleague Wilson da Silva. But Dr Finkel's wife is also a scientist and wanted additional information about what was involved when they signed up in October 2005. She was given a full briefing, which eased her anxieties.

'We were able to sit down and talk frankly about what was going to happen,' says Stephen Attenborough. 'This has been the great thing about selling the trips into space. You don't have to exaggerate the experience or avoid talking about certain things, because it all stacks up. Safety is the guiding star.'

Medical wellbeing is going to be a vital ingredient for the sustained success of this fledgling industry. Any early casualties will set the industry back years. Here the medical records of Greg Olsen, the fourth space tourist and someone found to have a smear on a lung X-ray, are invaluable. The level of detail is unprecedented and Olsen's experience has been very helpful to the field of aerospace medicine.

'He has been a fascinating guinea pig – willing to help advance the science. Choices are being made based on Greg Olsen's medical information,' says Chris Faranetta of Space Adventures. He agrees that a good bill of health for every passenger is critical. 'An interesting thing is that there is this ride at Walt Disney World called Mission to Space; more than twelve million people have been on that ride – two people have died. Essentially the two people who died – one was a child and the other an older woman – had cardiovascular problems. What I am saying is that if people have died on a Walt Disney ride, then we need to have our requirements in order and medical schools need to have their requirements in order. One of the issues is the cardiovascular and how normal people cope with two to three Gs, even if it is only for a short time.'

On the medical score, Dr Tizard is being supported by Wyle Laboratories, headquartered in Houston. They are a major life-sciences laboratory working closely with NASA and the US Department of Defense, and they have built up huge experience over 35 years, including clinical systems for the International Space Station and Mir. In July 2006, Vernon McDonald was named as director of the company's new Commercial Human Spaceflight Services unit, which will be focusing on space tourism. He has a doctorate in kinesiology – the scientific study of human movement – and oversees the company's space medicine unit at NASA's Johnson Space Center. Dr Tizard is also in discussions with the Johnson centre about how they can help too. They have nearly fifty years of flight experience to draw upon, having first selected astronauts in 1959. Since then nearly 350 US astronauts and 50 space explorers from other nations have trained at the centre. Here potential astronauts undergo one of the world's most competitive selection processes and it takes two years of intensive training before beginning special mission training.

Only then are professional astronauts eligible to be candidates for a flight.

Those astronauts have to learn about space-station systems and a variety of other disciplines including earth sciences, meteorology, space science and engineering. By comparison, a space tourist's training will be shoehorned into three or four days – but they will still have to undergo some of the basic astronaut training.

Part of this might involve a session in a large water tank. The Johnson Space Center houses the world's largest indoor pool, the Neutral Buoyancy Laboratory, which holds 6.2 m gallons of water and is more than 200 ft long and 40 ft deep. Within this pool, which simulates the weightless environment of space, astronauts train for space walks on full-size replicas of space-station modules. They spend ten hours in the water for every hour they spend walking in space. NASA is also teaming up with Wyle Laboratories to look at how humans will respond in a two-year weightless trip to Mars.

But first steps first. The stress points for the space tourist will be before and during takeoff, although landing in SpaceShipTwo will be much easier as it returns to land on the spaceport runway. Even a record-breaking NASA commander such as Mike Foale admits he still gets very scared. 'I don't like taking off and I don't like coming back down. But it's worth it for the experience in space,' he says.

Tizard sees the sudden acceleration after the hybrid-rocket motor is lit as being one of the most stressful moments of the flight – but also highly exhilarating. 'Obviously the G-force is the big one, but we don't think there will be an issue with weightlessness. This is generally a fun thing which people adapt to fairly quickly.

'Flying in a fast jet gives you a more realistic experience of what it is like to feel Gs in flight. Giving everyone an experience of this might well involve a spin doing some aerobatics,' she said.

Part of Virgin Galactic's space medicine programme will be a basic psychological assessment of each astronaut. 'Underlying all the training elements will be this psychological process to make sure that people will be relaxed and non-anxious about the flight. We don't want people to suffer panic attacks when they are seconds away from blast-off. So we will be giving them little pointers about how they can relax themselves.'

At 3 G, you can only move your arms and head under extreme pressure, but Tizard says that if your body is feeling the force of 3.5 to 4 G on launch there is nothing much you can do about it. 'You can't fight it. The force is too much. So you can't panic too much.'

Tizard calls up to 3 Gs the 'fun Gs' – but once it goes to 4 and beyond it becomes increasingly uncomfortable. 'You shouldn't try and move when you are faced with this pressure because you can strain yourself. And by the time our astronauts are up there in space and looking at the fantastic view – they will have overcome any ill-effects of the launch phase.

'The re-entry is just an added extra to the experience,' says the doctor. SpaceShipTwo's re-entry will take astronauts above 4 Gs for around 20 seconds, peaking at 6 G for no more than a couple of seconds, which will be bearable for most moderately fit people. 'I've done 4 G for thirty seconds and up to 6 Gs for thirty seconds. You feel incredibly heavy with six times your body weight on you.'

When it is front-to-back it puts pressure on the diaphragm and you have to breathe quite shallowly. On re-entry Virgin Galactic astronauts will be in their seats on the floor of the capsule – which will relieve the pressures.

But one regular question needs explanation: will space tourists have to wear a spacesuit? Typically, astronauts don't need to wear special clothing when they are living and working inside the space shuttle or International Space Station. Apart from the lack of gravity, the inside environment is pressurised and very similar to the Earth's. Astronauts work in comfort, wearing shirts, trousers and socks but no shoes. It's only when someone needs to go outside for a space walk that things get more complicated.

Humans sent out into space without a spacesuit would become unconscious in fifteen seconds, and suffer permanent brain damage in just four minutes. The temperatures on Earth may dip to 32°F (0°C) on a chilly day, but in space it can get down to −148°F (−100°C) or rise to 248°F (120°C). With those extremes of temperature, it is vital for astronauts to put on protective gear before they go out to work.

Virgin Galactic astronauts will wear a light flight suit – but a new kind of protective suit called a Bio-Suit is now being developed for longer space exploration. Professor Dava Newman, of the Massachusetts Institute of Technology, has been developing the Bio-Suit, which is radically different from the gas-filled spacesuits used for the Russian and NASA space programmes.

One of the earliest gas-pressure spacesuits – the Mercury M-20 – was adapted from high-altitude US Navy aircraft pressure suits with an inner layer of neoprene-coated nylon and an outer layer of nylon with aluminium attached. It was difficult for an astronaut to bend his knees or elbows. Improvements were made for the Gemini G4C EVA suit, which used Dacron and Teflon, allowing a gas-tight pressure bladder to control temperatures, while the Apollo A7L/B suit was much softer and allowed far more movement. The Orlan-M or Eagle spacesuit, with a gas-pressure bladder that acts like a shell, is still used by Russia's ISS astronauts, while the ISS Emu (Extravehicular Mobility Unit) is used for space-walking and features aluminium and fibreglass, with the limbs attached via airtight bearings.

None allow much movement on the surface. And, if humans are to explore the Moon and beyond, they will need to undertake far more extra-vehicular activities (EVAs). The Bio-Suit uses a system called MCP (Mechanical Counter-Pressure), which wraps and stretches around the body like a tight bandage. The Bio-Suit uses smart materials to become like a second skin that is light and helps mobility.

'Locomotion is a top priority of exploring planetary environments and performing useful work. Since Apollo only a few spacesuit concepts have been designed to provide locomotion and all current designs significantly hinder an astronaut's physical performance. The Bio-Suit proposes to augment human capabilities by coupling human and robotic abilities into a hybrid of the two,' says Professor Newman.

In time, the tight-fitting Bio-Suit, which generates its own electricity through piezoelectric boots while you walk, could well become an integral part of a Virgin Galactic astronaut's kit.

But what if something goes hideously wrong? Tizard and her colleagues are working with the designers to prepare instructions to deal with any emergency situations that might arise in the unlikely event of a complete systems failure.

'Burt Rutan has this whole idea of the spaceship being a lifeboat in itself. We are trying to avoid ever actually having to get out of the spaceship. So we don't have to do any of the egress issues at the moment. Even without power the spaceship will be able to glide back down to Earth manually,' explains Tizard.

There will be a range of safety instructions for the space traveller, from an emergency landing to the rocket motor being shut off. These will be rehearsed in the simulators and on flights with WhiteKnightTwo, which will have an identical interior to SpaceShipTwo, before the actual space flight.

So what about weightlessness? Astronauts can only experience weightlessness for about half a minute at a time onboard NASA's steep-climbing KC-135 aircraft. So a full five minutes in space will seem like a long time.

In orbit, the feeling of continuous free fall lasts for days. Zero gravity is a misnomer though, as there are very tiny gravity forces – about a millionth of a single G – so this in effect is micro-gravity instead. The suborbital space tourist will experience this marvel.

Astronaut Dan Barry described his surreal experience of drifting in space when he arrived on the International Space Station, then with only two modules, Unity and Zarya, in place, in 1999. 'I floated exactly to the centre of Unity, where I could not reach the walls, and got stranded in the middle of the room. It's not easy to get stranded – I had to have my friends help me get perfectly still. Once I was stationary, my brain remained convinced that I could somehow manoeuvre or kick myself over to the wall. But when I reached out an arm, my body moved back and my centre remained in the middle of the room. I instinctively tried moving fast, then slow, then bicycled my legs. None of it helped. I just had to wait for the air currents to drift me to the wall.' It's a topsy-turvy world that even Lewis Carroll's fertile imagination could never have imagined – where up is down and your whole body simply hangs in midair.

NASA astronaut Jeff Wisoff also offers some thoughts on his experience that will help the Virgin Galactic first-timer. He was talking about the launch of the space shuttle. 'Right before the engines cut off, you feel like you have a bear sitting on you. It's three Gs. Then at the moment of engine cut-off, you go from three Gs to zero Gs instantaneously. The bear jumps off your chest, and you see your seat belt float upwards, which is kind of cool.'

Another medical concern is people becoming sick. The sense of weightless elation can dissipate quickly. The longer-term space tourist will face a peril that is impossible to avoid – space sickness. The NASA boffins call it Space Adaptation Syndrome and more than half of all shuttle astronauts – and even highly experienced test pilots – feel queasy and nauseous. On the way to the Moon in Apollo 8, astronaut Frank Borman became ill due to motion sickness. The human body's inner ear, which gives us our sense of balance, takes time to adapt to the weightless atmosphere.

Steve Oswald, another shuttle commander, says: 'It seems that the people you think are going to be fine with space sickness are the ones who never are, and the ones you think are going to be throwing up are the ones who are doing fine.

'Sometimes guys are semi-Velcroed to the wall, throwing up, while the folk you least expected to be heroes are just chugging along, executing the plan.'

The advice for space tourists will be to move around as slowly and gingerly as possible and not to nod their head too much. But the feeling generally only lasts a day or so, and by the third or fourth day space tourists will find weightlessness a strange and wonderful experience.

'This should be fairly straightforward for the initial Virgin Galactic flights into suborbit,' says Tizard. 'We won't be in space long enough for it to properly kick in. Our flights should be fairly smooth because we are flying a straightforward parabola, whereas on the Zero G planes, or the Vomit Comet, the jerking up-and-down motion causes the problems,' she adds.

So your trip of a lifetime is nearly over. What happens when you glide back down to the spaceport? What will you be able to tell your friends and loved ones who are awaiting your return with a big party? Dr Tizard has some strong words of caution about coming back down to Earth. 'There is a really interesting aspect that we have discovered about such a dramatic human experience. When you come back down to the ground, absolutely the last thing you want is to be mobbed. As an astronaut, you have just had the most incredible experience of your life. You simply want to go away into a nice quiet room so that your body has time to reassess itself and calm down. We will be ensuring that this happens.'

Most people will need this downtime to recover their composure, because it is a huge adrenalin rush. It will be a bonding time between passengers and the pilot to talk about what they have actually achieved.

'Only then are you ready to go out and have the big party and the celebration,' she adds.

And as a civilian collecting your space wings – you are entitled to party.

22. THE UNIVERSE IS WAITING

The full Moon gazes down on a still November night, casting surreal and spooky shadows on the desert floor. A lone coyote howls from a mountain ledge in the distance. A cluster of star gazers – their 4 × 4s and RVs parked half a mile away – are wrapped up in thick fleeces, sipping coffee from flasks, with their heads facing up at the glorious astral array of twinkling lights. There are about a dozen people huddled under a selection of portable telescopes – and there is some excited chatter.

This is another wonderfully clear night to view the heavens. The prominent constellations of Cassiopeia, Cepheus, Draco, Pegasus, Andromeda and Cygnus are easily picked out and identified by those in the know. The blurry Pleiades constellation is stunning, while fizzing meteor showers can be seen through a medium-range telescope. Also visible are the astrologers' favourites of Pisces, Aries, Capricorn, Sagittarius and Taurus. Lower in the sky, to the south, are the planets: green-tinged Uranus, icy-cold Neptune, visible with a set of binoculars, and Pluto, a former planet now relegated from our solar system's list. The chat is friendly and informed as several members of the Astronomical Society of Las Cruces help a group of Virgin Galactic astronauts with their astral bearings at Upham, near to Spaceport America. This is one of their Dark Sky Observing meetings and a hidden gem of a gathering in a world infused with so much artificial light. The society was formed in 1951 by a dedicated band of astronomers including Clyde Tombaugh, who discovered Pluto. Once a month, its members pack their gear into their vehicles and drive away from the city lights to view the wonders of the universe through the pristine dark skies of southern New Mexico at 4,500 feet above sea level.

But the Moon is a dominant presence. So near, you feel as if you can reach out and touch the larger craters, its pock-marked expression etched by a thousand micro-meteorites bombarding its surface.

The awesome nature of space is easy to understand out here on this clear and cloudless New Mexican night. And there is a realisation that Destination Space is just beginning. The first Virgin Galactic astronauts will be flying just over sixty miles above the Earth into space. They will need to go another 238,790 miles to reach the Moon.

Space Adventures are now selling seats for a trip to the Moon. 'It is a $200 million programme,' says Chris Faranetta. 'We have two seats available at $100 million each. It's a three-person mission. It is based on some technology from the 1960s. A key technical component that we will utilise is the Zon programme, a Russian re-entry capsule that is very safe.'

Beyond this are the other planets of our solar system. Mars, the red planet, is the next stop. The illustrious astronomer Patrick Moore said in a Cambridge University lecture in 1961 that only Venus and Mars were in range for manned flight. But we now know Venus's surface temperature is intolerable, with carbon dioxide producing crushing pressure and clouds of thick sulphuric acid. So Mars appears the only option.

Mike Griffin, the head of NASA, has said: 'Beyond the Moon is Mars, robots first. Most internationals are at present more interested in Mars. Fine, we can't tell them what to be interested in. But our road to Mars goes through the Moon, and we should be able to enlist them to join on that path.'

Taking space tourists to Mars may well be decades away. Of all the planets in the solar system, Mars resembles the Earth most – but it too is a terrifyingly hostile environment, with surface storms and thick clouds of toxic hydrogen peroxide dust. The European Space Agency plans to send 'Bridget' – the ExoMars robotic research rover – in 2011, touching down on Mars in 2013. Its job will be to drill into Mars's surface to see if biological molecules exist under the surface. To borrow David Bowie's song title: is there 'Life on Mars'?

Another option might well be visiting Saturn's rainy moon, Titan, which has a landscape of mountains, hills and valleys, and lakes filled with methane and ethane, perhaps a future source of energy for a beleaguered Earth running out of its own natural resources.

On reflection, suborbital space for the mass market is a tiny, incremental step for mankind. But the inspirational designs of Burt Rutan and others must become a catalyst for a new generation of engineers, physicists and scientists. Space is a high-technology domain. For Scaled Composites, the next stage has to be orbital – and then perhaps a Moon mission.

Space belongs to everyone. And, as Sir Richard Branson says, it is still virgin territory. What we have discussed are tangible moves to take more tourists into space. Outer space is beckoning. And the Outer Space Treaty signed in 1967 means it must be used only for peaceful purposes.

The launching of Sputnik in October 1957 and intercontinental ballistic developments in rocketry led the United States to propose international verification of the testing of space objects. In September 1963, the USSR's foreign minister Andrei Gromyko told the UN General Assembly that the Soviet Union wanted an agreement banning the orbiting of spaceships or missiles carrying nuclear weapons. The United States replied, saying it had no intention of orbiting weapons of mass destruction, installing them on celestial bodies or stationing them in outer space.

The UN General Assembly unanimously adopted a resolution on 17 October 1963, welcoming a joint statement by the two superpowers and calling upon all states to refrain from introducing weapons of mass destruction into outer space. But it took more negotiations before a final treaty declared space an area of peace. The Treaty came into force on 10 October 1967.

Its most important point is Article Four. Firstly, it contains an undertaking not to place in orbit around the Earth, install on the Moon or any other celestial body, or otherwise station in outer space, nuclear or any other weapons of mass destruction. Secondly, it limits the use of the Moon and other celestial bodies exclusively to peaceful purposes and expressly prohibits their use for establishing military bases, installations or fortifications; testing weapons of any kind; or conducting military manoeuvres.

So, in spite of President Bush's recent determination to have America's military dominance in space, the future commitment to space has to be for peaceful means.

But what about distant galaxies. Will mankind ever be able to propel itself the vast distances required to visit them?

NASA's Voyager probes – Voyager 1 and Voyager 2 – have only recently reached the edge of our own solar system. Launched in August and September 1977, they have spent nearly *thirty years* travelling at the astounding speed of a million miles a day (61,200 kph), slingshot onto the gravitational pull of each planet, right out to the heliopause – the limit of the heliosphere – and the gravitational influence of our sun and its solar winds. Outside of our solar system, these tiny crafts – which have sent back amazing data as they passed Jupiter, Saturn, Uranus and Neptune – will enter interstellar space and the unknown. Beyond Neptune, these probes have discovered a myriad of minor planets in the Kuiper Belt, some larger than Pluto. The measurement of distance is Astronomical Units – the 93 million miles between Earth and the sun. Sent in different directions, Voyager 1 is now over 100AU from the sun, while Voyager 2 is around 90AU away.

The distances are almost impossible for us to imagine. If the Voyager probes are only now leaving our solar system, then what is there further out? This would take generations for human travellers to reach, unless some new form of high-speed propulsion was invented and became usable.

Our galaxy – the Milky Way – is part of a local galaxy of thirty others. The centre is about 3,000 light years away. Our closest neighbouring galaxy is Andromeda, visible with the naked eye from the New Mexican desert, which is 2.8 m light years away. The distance to 61 Cygni – the Swan – is a mere eleven light years. This means light emitted from this star takes eleven years to reach us travelling at the speed of light – 186,282 miles per second, or 670 m miles per hour. Mind-boggling distances.

Our own local group is ten light years in size. The European Space Agency's Hipparchos mission, from 1989 until 1993, measured over 118,000 stars that are nearly 500 light years from Earth. The centre of our galaxy is 500 light years away in the direction of Sagittarius, while Virgo is a cluster that is over 200 m light years away.

So the scale of this future challenge in space is monumental. To reach for the stars will require co-operation on an unprecedented global scale – using the brains and knowledge of everyone on Earth. Today, only Russia and the United States have the vast experience in the implementation of large manned space-flight projects, and the industrial infrastructure required for independently developing an interplanetary vehicle to put humans on Mars. This needs to change to involve all nations. And Europe, Australia, Japan and China all want to be part of this great expedition.

The experience of the International Space Station has demonstrated the feasibility and advisability of developing such large-scale space projects on the basis of the international co-operation. Humans will need to survive for long periods in the vacuum of space. Prolonged life in zero-gravity, with its muscle and bone wastage, and the dangers of solar radiation on humans, are only some of the known hazards. This co-operation can take different forms. The American have considerable experience in crew landing on and taking off from the lunar surface, as well as in landing robotic spacecraft on Mars. It is reasonable to assume that NASA will assume the prime role for developing the lander, one of the most critical elements of the mission.

Russia has accumulated considerable experience in building and operating space stations. The tasks performed by an interplanetary orbiter would be very similar to those of the habitable modules of space stations. Russia could be tasked with the development of such spacecraft. RSC Energia's concept is for a solar-powered space tug – an interplanetary vehicle that could belong to both the US and Russia – which would make use of gravitational trajectories to make return journeys to Mars.

Will Whitehorn, the Virgin Galactic president, sees the space tourist becoming an integral part of such future missions – involved with the scientific research. In the same way that explorers such as Charles Darwin sailed around the world as an independent naturalist on HMS *Beagle* between 1831 and 1836, so some intrepid space travellers will be prepared to travel to the unknown and the unknowable.

So this is a rare point in time. Perhaps Philippe Starck is right, and space travel is in the human DNA because humans have an unquenchable desire to explore. But, beyond suborbital excursion space, this will mean increasing risks, something Peter Diamandis says we must all accept and embrace.

For the generation who recall the Apollo missions, the new space renaissance gives them a chance to sample a trip into space – a dream that had been suppressed and denied to so many. Stephen Attenborough, the head of Virgin Galactic's astronaut relations, says, 'People have booked up primarily because they believe it is going to happen. They want to be a part of this by helping to make it happen. Everything has come together with Burt Rutan and Virgin Galactic. If this doesn't work then what will? There isn't going to be another opportunity like this in their lifetimes – and they want to back it to the hilt. The view of many of the Virgin astronauts is: if

I can lend my skills, my services, my knowledge or my money to make it happen, by buying a ticket, then it is worth it.

The next tiny footstep will be for tourists living in space. Perhaps, within the next two decades, we will be going to a Virgin Galactic space station 1,000 miles up or onto a revolving von Braun hotel in space – with its amazing views of our Earth. Further out are L4 and L5 – one of five Langrangian points in space – the exact points where the Earth's gravity and the Sun's gravity cancel each other out, allowing a spacecraft to orbit as if it was a planet. Here there is scope for another space station.

But government funding alone will never be enough to reach for the stars – it must involve private individuals and capital. This is what Sir Richard Branson and Will Whitehorn call Gaia capitalism, a kind of space-age entrepreneurship encompassing both environmental and astral opportunity. Gaia is a theory developed by UK scientist Dr James Lovelock, which argues that the planet is a unified, self-regulating organism and that the world's climate is self-regulating, but at intervals can create conditions that make the world inhospitable for human life. Increasing human activity and particularly carbon emissions appear to be threatening a natural climate equilibrium in which human life has flourished for the past 10,000 years. So there is a rapid need to look beyond Earth to find suitable resources for our long-term, sustainable existence. It's a stark prognosis, but space exploration is the only logical conclusion, unless an increasing population learns to consume less or differently.

This fits with the European Space Agency's mission to plot the stars. This too is called Gaia, and is due to be launched in 2010. It will conduct a census of one thousand million stars in our galaxy. It will monitor each of its target stars about a hundred times over a five-year period, precisely charting their distances, movements and changes in brightness. Gaia – or a Global Astrometric Interferometer for Astrophysics – is expected to discover hundreds of thousands of new celestial objects, such as extra solar planets and failed stars called brown dwarfs. Within our own solar system, Gaia should also identify tens of thousands of asteroids. So our consuming interest in space does not look like subsiding.

Professor Ann Karagozian, head of the Combustion Research Lab at the University of California in Los Angeles, points out that interest in outer space is increasing once again with the new generation of US undergraduates.

'My experience is that students are always interested and excited in moving into unknown, uncharted territory, and space continues to be a source of such excitement. Our undergraduate student enrolments in aerospace engineering at UCLA continue to climb, and it's one of the most competitive majors for admission on our campus.'

This is reflected elsewhere around the globe. The star-gazers in New Mexico are not alone in being fascinated by what they can see through their binoculars and telescopes. An emerging generation is ready to take on the challenges. Back in the early 1960s, President John F Kennedy called it the New Frontier. But the barrier came down at the frontier post. That gate is reopening and waiting for inquisitive and peaceful visitors from Earth. So let's keep exploring.

ACKNOWLEDGEMENTS

I am old enough to remember the Moon landings in 1969. My mother let me stay up late that night. Thanks, Mum. Buzz Aldrin was my hero, so meeting him during the course of this book was something special – and I believe the Moon landings actually happened. This fascinating book project came about when I was approached by Will Whitehorn of Virgin Galactic and KT Forster at Virgin Books. I'd like to thank them both for considering me.

I would like to thank those who have agreed to be interviewed: including Will, Alex Tai, Susan Newsam, George Whittinghill, Stephen Attenborough, Julia Tizard, all of Virgin Galactic. I have undertaken about 60 original interviews for this book and I'd like to thank all those people who have given their time and input. I'd also like to commend the Royal Aeronautical Society in London for arranging a superb one-day conference on space tourism in June 2006. The most amazing session was with Burt Rutan who was protected by a phalanx of PR people. I had to by-pass them all and drive across the Mojave desert to catch him off guard and he was a most willing and effusive interviewee. I also have to thank Brian Binnie and his lovely family. I have cribbed a lot of his own experiences for my chapter on suborbital space flight. I would also like to express my appreciation to Anousheh Ansari, and her colleagues from Space Adventures. She was very obliging in giving her time in New Mexico. I would also like to thank Rick Homans of New Mexico – and the people of New Mexico, who have been gracious and willing to help. Regards also to Peter Diamandis, a true hero in this story. And a very special thanks to Greg Allen, a former record producer and one-time paraplegic, who lives in Cutter, near Truth Or Consequences, New Mexico. I stumbled on his house in the desert when my car was stuck and he helped me dig it out of the muddy clay. Cheers, Greg. Sorry you have to move on to make way for the spaceport.

At Virgin Books, I would like to acknowledge KT Forster and Ed Faulkner, a very erudite and supportive commissioning editor. I'd also like to thank Virgin Atlantic for the support on the flights, especially the return from Las Vegas. Also a metaphorical bunch of flowers to Terri Razzell and Jackie McQuillan, Head of Human and Interplanetary Media Relations at Virgin Galactic. Thanks also to Sir Richard Branson for having the imagination to go for this project. Thanks too to Ned Abel Smith for his help with photographs.

And last, but by no means least, I'd like to thank my wife, Gail, and my children, Sam, Florrie, and Katie, who have put up with my spaced-out expression for a number of months.
Kenny Kemp
Edinburgh
January 2007

BIBLIOGRAPHY

There is a universe of excellent books and astonishing websites about space. I have used the NASA.com sites, Space.com and the excellent online encyclopaedia, Wikipedia in preparing and checking out this book. I am not yet a true space buff – perhaps that shows. Indeed any bloomers are entirely of my own making, but I have tried to be accurate and portray a picture of people who are passionate about space tourism.

Among the books I have used are: *First Man: The Life of Neil Armstrong*, by James Hansen, Simon & Schuster, 2005; *The Right Stuff*, by Tom Wolfe, Bantam Books, 1980; *Heroes in Space: From Gagarin to Challenger*, by Peter Bond, Blackwell, 1987; *Frontiers of Astronomy*, by Fred Hoyle, Heinemann, 1955; *Eyes on the Universe*, by Patrick Moore, Springer-Verlag, 1997; *Space Shuttle: The First 20 Years*, edited by Tony Reichhardt, Smithsonian Institute, Dorling Kindersley, 2002; *Chuck Yeager and the Bell X-1: Breaking the Sound Barrier*, by Dominick Pisano, Robert van der

Linden and Frank Winter, Smithsonian National Air and Space Museum, Harry Abrams, 2006; *The Rough Guide to Sci-fi Movies*, by John Scalzi, Rough Guides, 2005; *A Closer Look at Science Fiction*, by Anthony Thacker, Kingsway Publications, 2001; *The Space Tourist's Handbook*, by Eric Anderson, Quirk Books, 2005; *Chasing the Wind: The Autobiography of Steve Fossett*, by Steve Fossett and Will Hasley, Virgin Books, 2006; *Riding the Jetstream*, by John Christopher, John Murray, 2001; *Voyager*, by Jeana Yeager and Dick Rutan, The Adventure Library, 1987; *Yeager: An Autobiography*, by Chuck Yeager and Leo Yanos, Bantam Books, 1985; *From the Earth to the Moon*, by Jules Verne, 1865.

Other publications and periodicals used in this book were: *Wired, Flying* magazine, *Flight International, Flight Journal*, the *Sunday Post, The Times*, the *Daily Mail*, the *Las Cruces Sun-News*, the *Santa Fe New Mexican*, the *Sierra County Sentinel* and the *Wall Street Journal*.

INDEX